SEEDS OF SUSTAINABILITY

Seeds of Sustainability

Lessons from the Birthplace of the Green Revolution

Edited by
Pamela A. Matson

Washington | Covelo | London

Library of Congress Cataloging-in-Publication Data

Seeds of sustainability : lessons from the birthplace of the green revolution / edited by Pamela A. Matson. — 1st ed.
p. cm.
Includes bibliographical references and index.
ISBN-13: 978-1-59726-522-5 (cloth : alk. paper)
ISBN-10: 1-59726-522-5 (cloth : alk. paper)
ISBN-13: 978-1-59726-525-6 (pbk. : alk. paper)
ISBN-10: 1-59726-525-X (pbk. : alk. paper)
1. Sustainable agriculture—Mexico—Yaqui River Valley—Case studies. 2. Sustainability—Mexico—Yaqui River Valley—Case studies. 3. Green Revolution—Mexico—Yaqui River Valley—Case studies. I. Matson, P. A. (Pamela A.) II. Title: Lessons from the birthplace of the green revolution.
S451.7.S44 2011
631.5′8097217-dc23
2011014039

Printed on recycled, acid-free paper

Manufactured in the United States of America

10 9 8 7 6 5 4 3 2 1

Keywords: agriculture policy, sustainable farming, economic development, doubly green revolution, North American Free Trade Association (NAFTA), *ejiditarios,* International Maize and Wheat Improvement Center (CIMMYT), crop diversification, irrigation, pesticides, fertilizers

CONTENTS

TIMELINE FOR AGRICULTURAL DEVELOPMENT IN THE YAQUI VALLEY, 1890–2004

1890	The Ministry of Development grants to Carlos Conant the right to open irrigation channels on the margins of the Yaqui, Mayo, and Fuerte Rivers and to launch their colonization.
1891	Conant and US investors establish the Sonora and Sinaloa Irrigation Company to execute the contract for irrigation and colonization.
1900	The Sonora and Sinaloa Irrigation Company completes 39 km of channels from the Yaqui River.
1901	The Sonora and Sinaloa Company goes bankrupt.
1902	The Sonora and Sinaloa Irrigation Company and its shareholders reach an agreement on payments. The company pays its debt with land.
1903	Conant receives a new concession from the Ministry of Development to irrigate and colonize the Yaqui Valley.
1904	Conant begins the Companía de Irrigación del Valle de Yaqui to accomplish development work in the valley.
	The Sonora and Sinaloa Irrigation Company sells its rights to the Richardson Construction Company, a California-based land development company.
1907	Conant dies at the age of 63.
	The railroad reaches the point known as Esperanza Station, which, years later, gives rise to Cuidad Obregón.
1909	Richardson Construction Company forms the Companía Constructora Richardson and negotiates a new contract with the Ministry of Development regarding construction and colonization.
	David Richardson begins a new company, the Yaqui Land and Water Company, with an initial capitalization of US$15 million.
	Esperanza Station grows to a population of 450.

1911	Companía Constructora Richardson establishes an agricultural experiment station and publishes a crop calendar, including recommendations for 72 crops.
1913	Civil war extends into Sonora.
	Farmers in the Yaqui Valley plant 11,000 ha.
1914–17	Development in the Yaqui Valley is delayed because of the war.
1920	The government and the Yaqui people sign a peace agreement, concluding fifty years of intense warfare.
1925	Irrigated area increases from 15,000 to 37,000 ha.
1926	The government cancels the concession to the Companía Constructura Richardson and buys its shares of the Yaqui Land and Water Company, paying US$6 million. The company turns all of its shares over to the government development bank.
1927	The State of Sonora declares the creation of Cajeme County.
	Agricultural producers organize a research station.
1928	The National Bank of Agricultural Credit takes over the irrigation system and land.
1930	Cajeme County (including Cuidad Obregón) grows to a population of 12,000.
1936	The National Bank of Agricultural Credit assumes control of development but later transfers control to the National Irrigation Commission.
1937	The government applies the Land Tenure Law and expropriates private land to distribute among new *ejidatarios*. Altogether, 17,000 ha of irrigated land is distributed to *ejidatarios,* with 27,000 ha remaining in the private sector. An additional 34,000 ha of new land is also allocated to *ejidatarios*.
	The government launches the construction of the Angostura Dam.
1938	Farmers harvest 53,000 ha.
1940	Cajeme County grows to a population of 28,000.
1942	Angostura Dam is completed, adding an additional 60,000 ha of irrigated area.
1943	The government and the Rockefeller Foundation launch a collaborative agricultural research program, forerunner of the International Maize and Wheat Improvement Center (CIMMYT).

1950	Cajeme County grows to a population of 63,000.
1951	Wheat yields average 1.5 t/ha.
	The government creates the Yaqui Valley Irrigation District.
1953	Oviachic Dam is completed, adding 108,000 ha of irrigated area.
1955	Farmers in the Yaqui Valley harvest 210,000 ha of crops.
1960	Cajeme County grows to a population of 124,000.
1961	The government establishes the Instituto Nacional de Investigaciónes Agrícolas (INIA), the National Institute of Agricultural Research.
1963	Improved seed from CIMMYT is first released to producers.
	El Novillo Dam is completed, mostly for electricity generation, but also allowing total Yaqui Valley irrigated area to grow to 233,000 ha.
1964	100% of producers use improved seed.
1966	International Maize and Wheat Improvement Center (CIMMYT) is formally established as an international center.
1970	Cajeme County grows to a population of 183,000.
	Wheat yields average 3 t/ha.
1975	The government expropriates 34,000 ha of private irrigated land, which is transferred to new *ejidos*. The Yaqui Valley is thus divided among private landowners (41%), *ejiditario*s (55%), and colonists (4%).
1980	Cajeme County grows to a population of 256,000.
1990	Wheat yields average 5 t/ha.
1994	Whitefly invades the Yaqui Valley.
2000	Wheat yields average 6 t/ha.
1997–2004	Regional drought seriously depletes reservoir levels for the Yaqui Valley.

ACRONYMS

AOASS Asociación de Organismos de Agricultores del Sur de Sonora (Association of Producer Organizations of Southern Sonora)

ASERCA Apoyos y Servicios a la Comercialización Agropecuaria (Support and Services for Agricultural Marketing, Mexico)

BANCOMEXT Banco Nacional de Comercio Exterior (National Bank for Foreign Trade, Mexico)

BANRURAL Banco Nacional de Crédito Rural (National Rural Credit Bank, Mexico)

CGIAR Consultative Group on International Agricultural Research

CIMMYT Centro Internacional de Mejoramiento de Maíz y Trigo (International Maize and Wheat Improvement Center, Mexico)

CIRNO Centro de Investigación Regional del Noroeste (Research Center for the Northwest Region, Mexico); formerly CIANO (Centro de Investigación Agrícola del Noroeste)

CNA Comisión Nacional del Agua (National Water Commission, [CONAGUA] Mexico)

COFEPRIS Comisión Federal para la Protección contra Riesgos Sanitarios (The Federal Commission for Protection Against Health Risks)

CONASUPO Compañía Nacional de Subsistencias Populares (National Company of Basic Commodities, Mexico)

ENSO El Niño-Southern Oscillation

EPA Environmental Protection Agency (US)

EQIP Environmental Quality Incentives Program (US)

FERTIMEX Fertilizantes Mexicanos (Mexican Fertilizer Company, Mexico)

FIRA Fideicomisos Instituidos en Relación con la Agricultura en el Banco de Mexico (Trust Fund for Agriculture, Mexico)

FIRCO Fideicomisos de Riesgo Compartido (Trust Fund for Shared Risk, operated by the Ministry of Agriculture, Mexico)

FONAES Fondo Nacional de Apoyo para Empresas en Solidaridad (National Fund for Social Enterprises, operated by the Department of Social Development, Mexico)

GATT General Agreement on Tariffs and Trade

INEGI Instituto Nacional de Estadística y Geografía (National Institute of Statistics and Geography)

INIA Instituto Nacional de Investigaciónes Agrícolas (National Institute of Agricultural Research)

INIFAP Instituto Nacional de Investigaciónes Forestales, Agrícolas y Pecuarias (National Institute of Forestry, Agriculture, and Livestock Research)

ITSON Instituto Tecnológico de Sonora (Sonoran Institute of Technology)

NAFTA North American Free Trade Agreement

NARS National Agricultural Research System

OECD Organization for Economic Co-operation and Development (Paris, France)

OSHA Occupational Safety and Health Administration (US)

PAIS Programa Agrario Integral de Sonora (Integral Agrarian Program of Sonora, Mexico)

PEMEX Petróleos Mexicanos (Mexican Petroleum Company, Mexico)

PIEAES Patronato para la Investigación y Experimentación Agrícola del Estado de Sonora (Agricultural Research and Experimentation Board of the State of Sonora)

PROCAMPO Programa de Apoyos Directos al Campo (Direct Farmer Support Program, Mexico)

PROCEDE Program de Certificación de Derechos Ejidales y Titulación de Solares Urbanos (Program for the Certification of *Ejido* Land Rights and the Titling of Urban House Plots, Mexico)

SAGARPA Secretaría de Agricultura, Ganadería, Desarrollo Rural, Pesca y Alimentación (Ministry of Agriculture, Livestock, Rural Development, Fisheries, and Food, Mexico)

SEMARNAT Secretaría del Medio Ambiente y Recursos Naturales (Ministry of the Environment and Natural Resources, Mexico)

WTO World Trade Organization

PREFACE

This book is being completed in the second half of 2010, but the story it tells stretches from the early part of the 1990s until the late 2000s. It is a story about an agricultural region in the very early stages of what we hope will be a sustainability transition. It is also about the interdisciplinary team effort to which we devoted our time and energy for many years, an effort that has resulted in some important things—scientific discoveries, new management tools for crop production and water resource management that are now used well beyond the Yaqui Valley itself, new perspectives on how knowledge and action can be most effectively linked, new insights into interdisciplinary research and outreach, and new insights into sustainability transitions. It is also a story about frustrations—research that didn't get done, relationships that we struggled to develop and maintain, knowledge that we could not link with action. Despite those frustrations, this has been an exciting and useful project, and we hope the book communicates some of that.

The operational origins of the Yaqui Valley project trace to an event in 1992. During the summer of that year, Rosamond (Roz) Naylor (an economist) and I (a biogeochemist) hosted a two-week workshop at the Aspen Institute for Global Change. Each year, the Aspen Institute tackles interdisciplinary issues related to global environmental change, and among its goals is the hope of improving the interactions between natural and social scientists in order to address these issues. The focus of this particular workshop—at the interface of food, agriculture, and environmental issues—was an outgrowth of discussions at a biweekly forum of environmental faculty from all disciplines that Roz had begun at Stanford University three years earlier. The institute's twenty plus participants included Peter Vitousek, a Stanford biologist, Walter Falcon, a Stanford economist, and Donald Winkelmann, an economist who was then the director general of the International Maize and Wheat Improvement Center (CIMMYT) headquartered in Mexico. All were critical to making this project a reality. G. Philip Robertson, an ecosystem ecologist at Kellogg Biological Station (Michigan State University) with enormous experience in agricultural systems, was also a major and influential player in these discussions, and there were many other social and biophysical scientists as well.

Discussions at Aspen were provocative and wide ranging. After amusing and sometimes exasperating interchanges, for example, on the meanings of *productivity* in our different professional languages, we found common cause in sustainability issues related to nitrogen fertilizer and its agronomic and environmental effects, irrigation, and agricultural policy. We also found common cause in our focus on high-productivity agricultural systems of the type that now feed most of the planet's people, even though we realized that sustainability challenges might be more easily addressed in small, local, and even subsistence systems where decisions are less externally driven. Several research proposals were roughed out during the discussions at Aspen, but interestingly, the Yaqui Valley per se played no role in them.

The location issue was resolved in a second key meeting held at CIMMYT headquarters in Texcoco, Mexico. As a follow-up to Aspen, Winkelmann invited several of us to Mexico in 1993 to interact with members of his professional staff, especially those from the wheat program. CIMMYT was beginning to move more aggressively on environmental questions, and the head of the wheat program at the time, Anthony Fisher, felt that collaboration with Stanford might prove both scientifically and politically valuable. Soon the discussion moved to *who* and *where* questions. Those questions were answered simultaneously with the inclusion of Ivan Ortiz-Monasterio, a senior CIMMYT agronomist located in the Yaqui Valley at CIMMYT's premier wheat experiment station. In the years that followed, Ortiz-Monasterio became the lynchpin in the leadership trio, serving as both research coleader and our most critical boundary-spanning individual, linking our research community with the farmers and decision makers of the valley, who are his friends.

Those early discussions across disciplines and across nations were critical to launching this project, but they were leavened by other critical influences that helped the project evolve. Perhaps most important for me was my involvement as a board member of the National Research Council's (NRC) Board on Sustainable Development, starting in 1994 and extending through 2000, and then as a leader of the NRC Roundtable on Science and Technology for Sustainability through the 2000s. Those roles challenged me to see the world through the eyes of the many other member experts, all of whom represented different disciplines and walks of life related to sustainable development and environment. Most critical of those members were Robert Kates, geographer emeritus from Clark University and the cochair of the NRC board, and William Clark, the Harvey Brooks Professor of International Science, Public Policy, and Human Development at the Kennedy School at Harvard University. Their early engagement in, and encouragement of, the Yaqui research helped lay the groundwork for much of the most integrative research that followed. Indeed, Bill Clark played an instrumental role throughout this project, especially in the knowledge systems research and vulnerability research discussed in this book. At the same time, I think it is fair to say that the Yaqui Project, a real honest-to-goodness, on-the-ground project, brought real-life experiences and perspectives to ground the board and roundtable discussions, and influenced the work of subsequent projects and committees. Ultimately, I hope that it contributed to the emergence of the field of sustainability science and encouraged others to work within place-based, human-environment systems in a quest for sustainability.

Other influences were also critical. The perspectives of the International Geosphere-Biosphere (IGBP) and Human Dimensions of Global Change Programs (IHDP) underlay much of our early interest in greenhouse gases, agriculture, and land-use change, and NASA's land-use/land-cover change program funded some of the early research. Perspectives from the Consultative Group on International Agricultural Research (CGIAR), of which Wally Falcon was a board member, influenced our perspectives on food security, food production, and sustainable agriculture. Roz Naylor and Ivan Ortiz-Monasterio brought their expertise on different dimensions of those perspectives as well as their enthusiasm for interdisciplinary problem solving. CIMMYT and its leaders made it possible to do the work.

For me personally, my husband and Stanford professor Peter Vitousek was a continuing and most important influence. He is an ecosystem ecologist and a global change scientist who has an uncanny ability to see the whole system and to identify the really important things to be done. He helped launch this project, advised throughout, read many manuscript drafts (including parts of this book), and with our kids, Mat and Liana, held down the fort at home during my many trips to the Yaqui Valley. Wally Falcon was a research team member but so much more—his savvy in the international agricultural world made many things possible, his wisdom kept us on track, and his writing abilities helped us to both find funding and get the word out. In addition to the authors of this book, Stephen Gorelick, a hydrologist, Karen Seto, a geographer, and Steven Monismith, a physical oceanographer, all professors at Stanford, brought their knowledge and expertise and ideas and great students to the project, as did Tracy Benning, Greg Asner, and many other scientist friends. José Luis Minjares of the Mexican Water Commission and Carlos Valdes-Casillas, who was at the Monterery Technical Institute in our early years, brought management reality to much that we did, and José Luis, along with Ivan, were essential linkers of knowledge and action.

Of course, there were many other people from Mexico and the United States who made the project work; the individual chapters of this book include acknowledgments of them. My personal thanks go to Peter Jewett, who managed the Matson lab through much of the research and has played a critical role in getting this book to press, and to Tina Billow, who managed in the start-up phase and in the "buried-in-data" stage, keeping multiple projects going in the field and lab. Lori McVay was our administrative leader throughout, and Ashley Dean, our research coordinator for much of the project; without them we could not have grown and expanded in the way we did. Mary Smith helped make our meetings happen efficiently and with great fun. These, our students and post docs, and countless other people made the Yaqui Project vibrant, and its successes are thanks to them. Finally, a great many funding sources made this work possible; they are listed in individual chapter acknowledgments, and include the Packard Foundation, the USDA, NASA, NOAA, and NSF, the Ford Foundation, the Pew Charitable Trusts, the Hewlett Foundation, the Andrew Mellon Foundation, and the MacArthur Foundation (whose generous fellowship to me helped give me the freedom to start down this road). The Packard Foundation provided the critical support that allowed us to develop an interdisciplinary effort rather than a project with many disciplinary pieces, and it also supported the synthesis effort that resulted in this book as well as other outreach products.

We hope that the contributions discussed in this book are lasting ones, both to the Yaqui Valley and, more broadly, to the field of sustainability science.

Pamela Matson

PART I

The Birthplace of the Green Revolution

Chapter 1

Why the Yaqui Valley? An Introduction

PAMELA MATSON AND WALTER FALCON

There are few agricultural regions in the world more interesting and important than the Yaqui Valley in Sonora, Mexico. The Yaqui River Basin has supported agriculture for many centuries, but the story we focus on is modern. The valley is the birthplace of the *green revolution*, and it is now one of the most intensive agricultural regions of the world, using irrigation water, fertilizers, constantly improving cultivars, and other inputs to produce some of the highest yields of wheat anywhere. It is one of Mexico's main breadbaskets and also a global supplier of seeds and grain. In this, the Yaqui Valley provides a story of agricultural and economic development that is emulated and reflected the world over. But over the past several decades, its story has also become one of environmental, resource, economic, and social challenges related to water resources, air and water pollution, impacts of global environmental and policy changes, human health concerns, biodiversity conservation, and climate change. As these two story lines have merged, this region has had to evolve and change. It is this story of early steps in a sustainability transition—a transition at the interface of environment and development—in which we engaged through our integrative research and outreach. It is this story that we hope to tell in this book.

Sustainability is a complex concept, one with multiple definitions and goals. In its report titled *Our Common Journey,* the National Research Council (NRC 1999) defined sustainability broadly as the goal of meeting the needs of people today and in the future while (and by) protecting the life support systems of the planet. As we use it here, in the context of transitions in the Yaqui Valley, we encompass the goals of improving and enhancing food, fiber, and potentially even biofuel production, protecting the economic and social welfare of the people of the region, and sustaining its resource base and environment on land and in the sea.

Worldwide, the sustainability challenges of agriculture and food security are enormous, given the need to feed a still-growing human population that is likely to plateau at near nine billion by the middle of this century. Today, in 2010, scientific concern about this challenge can be seen in the pages and special issues of *Science* and *Nature* magazines, among many other venues. Taken together, the growing food demand associated with population growth; alleviation of hunger and increased consumption of meat and dairy; the increasing competition of agricultural lands for other uses; the increasingly clear evidence that agriculture drives negative environmental and human health changes at local to global scales; and the growing, serious concern about the effects of climate change on crop systems have called for a worldwide research effort to address agricultural sustainability and food security (for recent reviews and analyses of these issues, see IASSTD 2009; Royal Society 2009; Godfray et al. 2010; Federoff et al. 2010; NRC 2010a, to name just a few).

Agricultural sustainability challenges potentially can be addressed through a variety of approaches, including, for example, new breeding technologies, including the development of genetically modified crops; new kinds of integrative crop-livestock systems; precision agriculture (both "high tech" remote sensing and computer-based mechanized approaches as recently described in Gebbers and Adamchuk [2010] as well as lower tech approaches that use information to increase input efficiency); agroecosystem approaches that seek to use soil, water, and light resources to maximize and increase efficiency of production while reducing environmental negatives; and new, more efficient aquacultural systems. While all of these can contribute to sustainability goals, none can do the job everywhere nor be implemented overnight to achieve sustainability. In the Yaqui Valley as in most agricultural systems, sustainability is not an end point that can be defined or that is likely to be achieved in the near term, but rather a process of developing options and making choices that increasingly honor these multiple goals, and that make progress toward all of them. The Yaqui Valley is still in the early phases of its transition to sustainability, but these first steps are important.

Our story is about these seeds of a sustainability transition in an agricultural region, but the things we've learned—about implementing win-

win opportunities, or knowledge systems for sustainable development, or vulnerability analyses of human-environment systems, for example—are relevant to many other sustainability efforts outside of agriculture. Likewise, what we have learned about the role and contributions of multi- and interdisciplinary research in developing options and supporting implementation of them speaks to sustainability science and development efforts more generally.

The Story of Agriculture in the Yaqui Valley

In our research in the Yaqui Valley, our primary focus was the dynamic human-environment systems in irrigated agriculture and nearby land and ocean systems, as they functioned between 1993 and 2008 at the end of the green revolution and the beginning of what Gordon Conway calls the "doubly green revolution" (Conway 1997). The longer story of agriculture in the valley is, however, of importance to the more recent past; chapter 2 provides a detailed history, but an abbreviated version will be useful in this introduction to the book.

Located on the northwest coast of mainland Mexico, bound by the Gulf of California to the west and the Sierra Madre Occidental foothills to the north and east (fig. 1.1), this productive coastal plain has been inhabited for thousands of years by the indigenous Yaqui Amerindians. For centuries, the Yaquis fought against and ultimately lost to the encroachment of Spanish and Mexican colonists interested in their fertile lands and silver resources. The introduction of foreign investment and irrigation in the 1890s and early 1900s laid the foundation for what would be the most intensively irrigated agricultural land in Mexico that now covers 233,000 hectares. The establishment in the 1930s (and thereafter) of a substantial number of *ejido* (collective) farming units, in addition to private landowners, the Yaqui Amerindians, and foreign investors, made for unusually diverse groups and interests within the region.

In the mid-twentieth century, the Mexican government and the international development community identified the Yaqui Valley as an appropriate center for agricultural research and development, given that it is agroclimatically representative of about 40 percent of wheat growing areas in developing countries. Led by Norman Borlaug and an international team of scientists, the wheat research program promoted intensive technologies such as new high-yielding crop varieties, large-scale irrigation, fertilization, and pesticides. The results—a dramatic increase in grain production that supported Mexico's transition to self-sufficiency in wheat production and

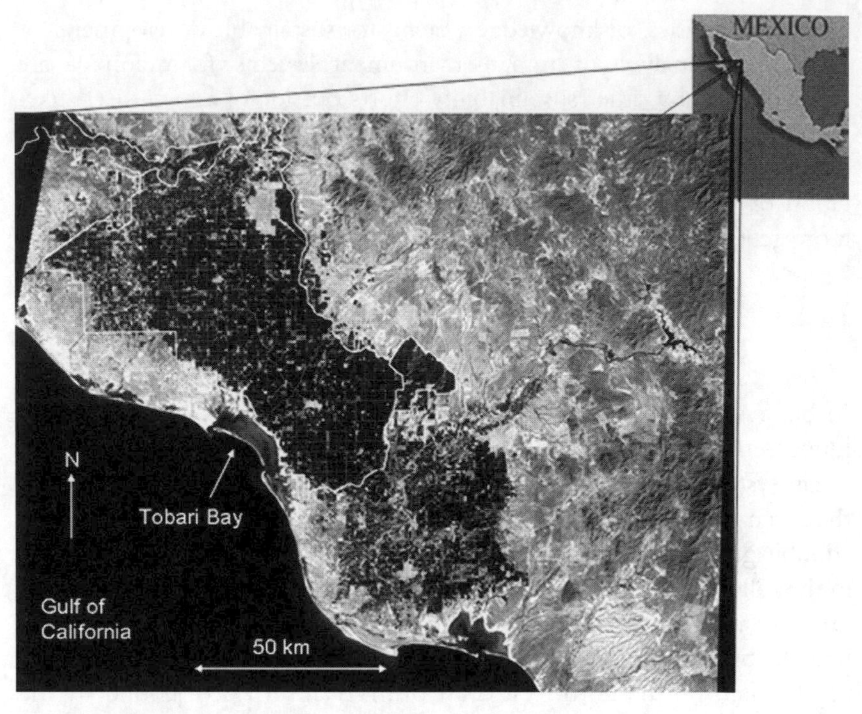

FIGURE 1.1 Location of the Yaqui Valley, Sonora, Mexico

the direct transfer of semidwarf wheat technology to South Asia in the late 1960s—gave the valley its recognition as the home of the green revolution for wheat. However, agricultural development in the region was not the only story of change. Rapid population growth focused in cities, major development of fisheries, the engagement of the international conservation community in the adjacent oceans, the development of coastal aquaculture, and the rapid increase in livestock operations were likewise a part of the story in the valley and surrounding regions. By the time our research team entered the picture, these changes were under way.

The period of the 1990s and early 2000s, however, involved further changes, and many of them were especially challenging for Yaqui citizens, especially farmers. An eight-year drought (1997–2004), coupled with questionable irrigation procedures, literally drained the valley's irrigation reservoirs dry, and raised questions about vulnerability of water resources in the context of future climate changes. Fertilizer use increased, but so too did nitrogen losses in the form of greenhouse gas and air pollutant emissions to the atmosphere as well as water pollutants. Crop diseases and pests came, but rarely went, causing for example the complete loss of soybeans from cropping rotations. Major changes in Mexican macroeconomic policy, Mexico's entry into the North American Free Trade Association (NAFTA), and booms and busts in international commodity markets created new forms of economic uncertainty in a valley that had previously led a very "policy-protected" life.

Constitutional changes expanded the ways in which *ejiditarios* (small communal farmers) could rent and sell their land, but also made them more vulnerable to other market-oriented policies on credit and fertilizer. Many aspects of the irrigation system were decentralized from federal to state and valley organizations, giving local farmers more authority, but also more responsibility for the ways in which water systems were managed. Agricultural extension shifted from federal hands to those of farmer unions. Attempts at diversification into fruits, vegetables, livestock, and aquaculture solved some problems, but created other ecological and economic dilemmas in the process.

These physical, economic, environmental, and social changes greatly complicated life in the Yaqui Valley. They also complicated our research efforts, but at the same time made those efforts more interesting and valuable. Change was happening so rapidly that Yaqui residents eagerly sought out the results of our studies, but the fast pace also made it difficult to establish sensible research priorities, let alone to fund them. For all of its limitations, however, the research program reported on in this volume was quite remarkable. Much place-based research is based on a single snapshot in time. We do not have a complete historical movie of the Yaqui Valley, but we do have a fairly complete ten-plus year video clip.

Integrative Research in the Valley

The choice of the Yaqui Valley as the focus of our research effort was in some ways just good luck (see the preface for more on the origins of the project), but after the fact, it proved to be an excellent choice. It is a region small enough to be understood, yet large enough to be interesting. It is a region that represents the kind of high-productivity, surplus agricultural system that is key to feeding billions of people; indeed, we explicitly chose this kind of system, in contrast to small scale, localized, or subsistence systems, even though sustainability decisions in the valley were likely to be more complicated and externally controlled. It is a region that makes connectivity—between land and ocean, land and air, water and food, country and country, peso and dollar—very obvious.

The Yaqui Valley was also an excellent choice for our research because it has been a focus of disciplinary research, mostly agronomic, for decades. As the location of the primary field station of the International Maize and Wheat Improvement Center (CIMMYT), one of the major centers of the Consultative Group on International Agricultural Research (CGIAR) system, it has accumulated more than thirty years of knowledge. CIMMYT field research and survey data on valley farmers and on wheat technology provided a wonderful base from which to build our research. The federal government has also been conducting agricultural research through the National Institute of Forestry, Agriculture, and Livestock Research (INIFAP). INIFAP operates eight regional agricultural research centers throughout Mexico, including one in the Yaqui Valley (Research Center for the Northwest Region, Mexico [CIRNO]). CIRNO works closely with CIMMYT to further research on genetic improvement, production technology, pest management, new cropping options, and irrigation technology. Also, the farmer-owned Agricultural Research and Experimentation Board of the State of Sonora (PIEAES, known locally as the *Patronato*) receives breeder quality lines from CIRNO and sells them to farmer organizations who produce registered seeds. And perhaps the most important actors engaged in agriculture research are the farmers themselves—particularly a set of innovative farmers that routinely collaborate with CIMMYT and CIRNO scientists—and the credit organizations that represent them. The National Water Commission (CNA) also carries out research in support of water for irrigation, and universities such as the Sonoran Institute of Technology (ITSON) and the Monterey Technological Institute conduct research in agronomics and related areas, especially natural resources and coastal zone management.

With all this research strength, one might wonder what our multidisciplinary team of researchers from the United States and Mexico could offer. The answer, quite simply, was an integrative systems perspective. We started with a focus on the human-environment systems of the place, and an interest at the interface of crop yields, economic gains, fertilizer use, and environmental implications. Before long, as our understanding of the valley's challenges became clear and as our team grew, we focused on irrigation management, aquaculture development, Mexican and international economic and agricultural policies as they affected the valley, environmental links between the irrigated valley and the coast and the Gulf of California, diversification of crops, and climate change and vulnerability in the agricultural sector, and others. It did not take long to see that many if not all of these issues were connected, and that the Yaqui Valley was an excellent place to analyze and understand them as a system, perhaps doing something to help manage them sustainably. We did not start with the intention of developing a broad, decade-long analysis of sustainability in the valley, or of engaging with decision makers across such a broad range of issues, but our research took us there. Over time we saw that there is no better place to evaluate where progressive agricultural systems were headed technologically; to view the effects of globalization and the impact of NAFTA; to make the connections between agriculture, resource use, and environmental impacts; to engage in efforts to share new knowledge in decision making; and to understand those knowledge-to-action linkages, than in the Yaqui Valley.

This work was completed through the support of many different projects by many different funding organizations (see the preface for details). The need for multiple funding sources complicated the knowledge-generation process. Some critical pieces were never funded, frustrating our desire to understand a more complete story and provide more useful information. The time perspective and the long-term nature of the inquiry were crucial, but also called for an almost constant search for funding. Learning and modification of ideas took place in laboratories, at the experimentation station, on farmers' fields, in scores of meetings with growers and other decision makers, and in regular team meetings both in the Yaqui Valley and at Stanford University. While the research elements at times appeared disjointed, they emerged logically and often built upon each other as they progressed. The individual subprojects were well done, and their outcomes have stood alone as published research papers as well as new models, tools, and management approaches. While it may be that these individual products yielding from these projects have mattered most to decision making in the valley, it is the sum total that best tells the story of the valley in transition.

In subtext, this book also tells the story of a changing and growing team of researchers trying to move from simply understanding the challenges being faced by the people and ecosystems of the valley to assisting in addressing those challenges. How, over time, did agronomists, biogeochemists, economists, ecologists, engineers, oceanographers, geographers, hydrologists, and other scientists decide to work together to analyze and help solve problems? How did we interact with farmers and other decision makers of the valley, and how did the flow of information among us all determine which problems were chosen and why? How was the research funded as problems and funding agendas changed? And what research lessons were learned, both positive and negative, from a research effort that covered more than a decade and cost several millions of dollars?

The relatively long-term extent of the project proved the importance of working in one place continuously, and for a long time; had the inquiry stopped in 1998, or even 2002, our understanding of the interface between economics and environmental issues of the valley would have been substantially more limited. Nevertheless, it is clear that the full story of the Yaqui Valley is not ours to grasp. The valley continues to change, and some, perhaps much, of what we learned during our years of joint study and engagement has become outdated. We seek, therefore, to share through this book some of the general learning and more generally useful information, perspectives, and research approaches, along with the specific knowledge that was useful at a given time and place.

Organization of the Book

We tell the story of the last few decades of agricultural development and environment in the Yaqui Valley in several parts. In this first section, we introduce the reasons for working on these issues and in this valley, and, in chapter 2, we set the valley in historical perspective. Then, in the second part—chapters 3 through 7—we tell the interdisciplinary, integrative stories that motivated much of our work. We asked, for example, whether win-win-win solutions, for economics, agronomics, and environment, are possible in the wheat fields of the Yaqui Valley, and what would be needed to make them a reality (chap. 3). This is the project that got us started, and it eventually led to new management approaches that are perhaps first steps for this valley in terms of sustainability transitions. However, this study's initial contribution was to show how research teams of economists, agronomists, and environmental scientists, working together, could advance science and also support decision makers. We tested the idea that intensive agriculture "spares land for nature"—that is, that carrying out agriculture very intensively (with high productivity) in some places allows and leads to less additional conversion of natural ecosystems to agriculture (chap. 4). We found that, at least when evaluated at the regional scale, intensification has direct causal links to off-site activities that have negative impacts on natural systems, and concluded that, if one is worrying about sparing land for nature, intensification is necessary but not sufficient. One has to protect land for nature purposefully, not as an assumed by-product of intensification.

We studied and evaluated the knowledge system of the Yaqui Valley, trying to understand the most critical dynamics by which new knowledge is linked to decision making, and in doing so, learned much about our own role in the knowledge system (chap. 5). We built on our knowledge of the human-environment system of the Yaqui Valley agriculture and water sectors to develop approaches by which to evaluate vulnerability (the likelihood of people and ecosystems suffering harm) to forces such as climate changes or policy changes (chap. 6). Finally, we explored the links between land and sea, including the drivers and consequences of rapid aquacultural development as well as agricultural intensification (chap. 7). These integrative, interdisciplinary studies helped us develop an understanding of the valley that went beyond what was happening to cropping systems and the producers that managed them. They also provided data and perspectives that were then generalized through comparative analyses with other places around the globe. Indeed, the development of the Yaqui Valley as a case study has had implications for understanding and managing human-environment systems that extend well beyond the valley margins.

The third major part of the book, chapters 8 through 11, provides some of the deep knowledge base that was developed about components of the Yaqui Valley system and underlies the interdisciplinary analyses about the Yaqui human-environment systems. Issues such as macroagricultural policies (chap. 8), crop management systems (chap. 9), fertilizer use and nitrogen flows from land to the sea (chap. 10), and water resource development and management (chap. 11) are dealt with in detail, representing many years of research and the involvement of many researchers; they also provide the building blocks with which the most interdisciplinary questions could be addressed. Although the chapters in this third part present research and new knowledge that cuts across sectors or institutions or landscapes and are thus themselves integrative in sometimes unique ways, they tend to be more narrowly and more disciplinarily focused than the chapters in the second part. Thus, they will be useful to students and researchers who are pursuing disciplinary analyses (for example, in agricultural policy, biogeochemistry, agronomy, or water resources research) as well as those who are interdisciplinary.

The final part of the book—chapter 12—provides a brief, retrospective glance at what we have learned, and failed to learn, in the project overall. In this chapter, we also reflect on our role as researchers in the knowledge system of the Yaqui Valley, the challenges of interdisciplinary research, and the successes, and more often the inadequacies, of our attempts to link knowledge to action for sustainability transitions in the valley.

All but one of these chapters is authored by two or more members of our research team. These author lists indicate the lead members of the research teams involved in our studies, but they are not inclusive of every participant in the research. Indeed, it is fair to say that every author of this book was involved, in one way or another, in every aspect of the research described here, and many important players are not listed as authors at all. Acknowledgments in each chapter attempt to recognize the many contributors, but we no doubt have missed some.

We have endeavored in this book to tell a coherent story of our substantive findings. To do so, we chose to write new chapters from scratch, rather than simply compiling previously published articles. Each of these chapters is meant to contribute to the broad, overarching view of transitions in the Yaqui Valley, but each can also be read alone, depending on the reader's interest. Whether the reader chooses to read part or all, we hope that our experiences in the valley provide ideas, insights, and knowledge that will be useful to those interested in transitions to sustainability.

Chapter 2

A Brief History of the Yaqui Valley

ASHLEY DEAN

The history of the Yaqui Valley reveals an intricate story of a place constantly facing and adjusting to change over time. From its management for thousands of years by the Yaqui Amerindians to its current state as a highly productive breadbasket, this rich coastal plain has provided food and other essentials for its inhabitants, has been contested ground among many different groups, and, more recently, has become an interesting model of agricultural development for the developing world. This short history attempts to trace those transitions to the present. Many of the issues raised here, especially in the most recent history, are the focus of the chapters to come.

Early History: 500s–1930s

Records dating back to AD 552 indicate that the Yaqui Valley was actively inhabited by Yaqui Amerindians living along the Yaqui River. The Yaquis supported themselves by hunting, gathering, and fishing in the productive marine ecosystem as well as through cultivation of corn, beans, and squash. They traded such goods as native foods, furs, shells, and salt with other indigenous groups in Central America and were known to roam extensively during pre-Columbian times. Beginning in the fifteenth century, the Yaquis organized into autonomous, yet unified, cultural and military groups in order to defend themselves from repeated attacks by the Spanish military over the next couple of centuries.

While initially successful at defending their land and cultural integrity, they were not a warring tribe. They welcomed Jesuits into their villages in the early 1600s to do missionary work and to support economic development. Most of the 60,000 Yaquis settled into eight *pueblos*, each centered around the community church where a unique belief system was created, blending Jesuit teachings and ancient beliefs; the beloved deer dancer remains a prominent religious figure within Yaqui culture today.

The discovery of silver in the Yaqui Valley in 1684 disrupted the peace once again as Spanish settlers moved in and aggressively took over Yaqui land. In an effort to preserve their lands, the Yaqui Indians allied with the neighboring Mayo tribe in 1740 to force the Spanish out of their territory. For the next 190 years, the Yaqui people continued to fight both the Spanish and the newly independent Mexicans, suffering greatly in the process. Some escaped to the Vakatetteve Mountains, while others relocated to Yaqui communities in Arizona. Those who stayed continued to fight for the right to their land and autonomy. By the late 1800s, violent encroachment and the spread of smallpox had reduced the Yaqui Amerindian population in the Yaqui Valley to 4,000. The community was further fragmented by Sonoran policies that supported the deportation of both peaceful and rebel Yaquis to Arizona communities to make way for the foreign investment movement that marked the beginning of the 1900s.

In 1890, the Ministry of Development under President Porfirio Díaz identified the Yaqui Valley as a promising area for colonization and agricultural development. Soils of the Yaqui Valley were recognized as particularly fertile—rich in alluvium and lime—supporting the production of a variety of crops. Located 300 miles south of the US border, the valley also promised to be a competitive trade partner in international markets, in addition to accommodating strong domestic market demand.

Although the climate supports a year-round growing season, developers had to contend with an average precipitation rate of only 30 cm of rain annually. The seasonal distribution of precipitation in the Yaqui Valley is dominated by the North American monsoon with over two-thirds of the precipitation occurring from July to September (fig. 2.1). Variability of the climate, high evapotranspiration rates (200 cm are potentially lost each year) and minimal precipitation during the main growing season (November–April) necessitated the development of an expansive irrigation network (Addams 2004). Fortunately for developers with access to both the Yaqui and Mayo Rivers, the projected area for irrigated, agricultural expansion in the Yaqui Valley was over 300,000 ha.

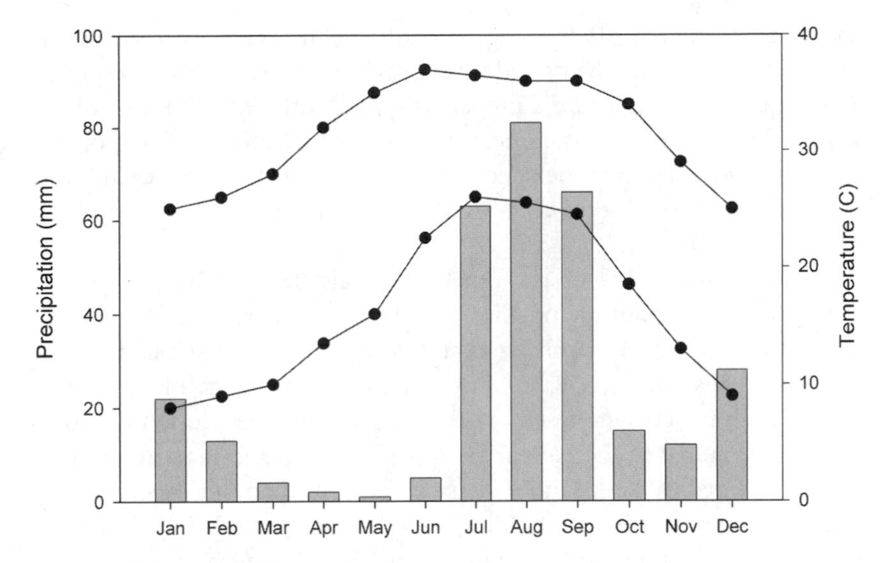

FIGURE 2.1 Average monthly rainfall and minimum and maximum temperatures in the Yaqui Valley, computed from daily measurements at CIMMYT station for 1969–2003.

Díaz authorized a program of colonization financed by American banks to divide the Yaqui Valley into parcels and to build the necessary network of irrigation canals and reservoirs (Salinas-Zavala et al. 2006). The government granted developer Carlos Conant the right to colonize 300,000 hectares of land and provided concessions for water from the Yaqui, Mayo (Sonora), and Fuerte (Sinaloa) Rivers to open irrigation channels. In 1891, Conant, through the Sonora and Sinaloa Irrigation Company, built the Canal Bajo and by 1900 had completed 39 km of canals off the main channel in the Yaqui River.

In 1904, the Sonora and Sinaloa Irrigation Company sold its rights to the Richardson Construction Company, a California-based land development company. David Richardson, main shareholder of the Richardson Construction Company, established the Yaqui Land and Water Company in 1909, and continued the construction of irrigation infrastructure and colonization in the Yaqui Valley (Salinas-Zavala et al. 2006). The company attracted investors—advertising the Yaqui Valley as "The Most Fertile Irrigated Land on Earth"—and sold an acre, including water rights, for US$25 (Yaqui Land and Water Company 1909). Promotional material promised "a better combination of soil, water, markets, transportation facilities, products and climate than any other tract of irrigated land ever offered in North America. . . a chance to make a fortune in few years" (Yaqui Land

and Water Company 1909). Corn, wheat, garbanzo beans, cotton, sugarcane, alfalfa, and other fruits and vegetables were touted as chief products. The Yaqui Land and Water Company supported the development of the valley's first agricultural experiment station and published the first crop calendar for the valley that included recommendations for over seventy-two crops. Further expansion was delayed in 1910 with the beginning of the Mexican Revolution.

The revolution began as a rebellion against Porfirio Díaz's dictatorship and land distribution inequities. At the time, 1 percent of the population owned 97 percent of the land and 96 percent of the population was landless (Lewis 2002). It was a bloody and violent war, and many families moved out of the valley to escape the violence. The revolution ended in 1917 and resulted in the *Plan de Ayala* which mandated that one-third of all lands owned by large landholders be distributed among landless peasants grouped into collective holdings called *ejidos* (Lewis 2002). Article 27 of the 1917 Mexican Constitution established the *ejido* land reform program, and declared all land and water ultimately the property of the nation. While private property was allowed in principle under the new *ejido* system, Article 27 established a legal limit on private land holdings of 100 irrigated hectares or the "non-irrigated equivalent."[1] It also fixed the size of the *ejidatario* parcel at a minimum of 10 hectares of irrigated land (Lewis 2002). Furthermore, Article 27 mandated that the state reserved the right to expropriate private land holdings when they exceeded the legal limit and to reclaim *ejido* lands when they were improperly managed.

With the end of the revolution, conditions stabilized within the Yaqui Valley and the irrigated area increased to 37,000 hectares by 1926. That same year the government canceled Richardson's construction contract and paid US$6 million to turn all its shares of the Yaqui Land and Water Company over to the government development bank, *Nacional Financeria*. The National Bank of Agricultural Credit took over all irrigation systems and lands in the Yaqui Valley, strengthening and increasing the infrastructure to open new areas for cultivation (Salinas-Zavala et al. 2006). It marked the beginning of a period favoring the nationalization of resources and land redistribution, and a move away from a reliance on foreign investment.

The first major land distribution in southern Sonora occurred in the 1930s under President Lázaro Cárdenas. On a national scale, the Cárdenas administration redistributed 20 million hectares to approximately 800,000 people over a six-year period (1934–40), decreasing the number of landless laborers in Mexico by over 50 percent (Lewis 2002). In the Yaqui Valley,

Cárdenas distributed 36,000 hectares among 2,160 peasants organized into fourteen *ejidos*. The majority of the redistributed land was managed collectively for grazing and rain-fed agriculture, while 17,000 hectares were irrigated (Dabdoub 1980). In 1939, President Cárdenas granted the Yaqui tribe official recognition and title to its land—accounting for 25,000 hectares of agricultural land in the Yaqui Valley—ending centuries of violent dispute. Several larger parcels of coastal land were also granted to these indigenous communities (Luers et al. 2006).

Research, Development, and Intensification: 1940–1989

The 1940s through the 1980s were years of rapid change, marked by irrigation expansion and agricultural intensification in the Yaqui Valley, as well as a range of changes in land ownership and tenure.

The Green Revolution

In 1942, the Angostura Dam was completed, adding 60,000 hectares of irrigated area to already over 53,000 hectares of harvested land. Several years later, the Oviachic and El Novillo Dams were built along with a new canal, the Canal Alto, permitting the irrigated area to grow to its present level of 233,000 hectares (Salinas-Zavala et al. 2006).[2] The irrigation system includes over 2,700 km of main, lateral, and secondary irrigation canals, delivering over 2 billion m^3 of reservoir water each year to the valley's farmers (McCullough 2005).[3] Local irrigation districts N° 18 (the oldest in northwestern Mexico) and N° 41 continue to control the canals and are responsible for releasing reservoir water directly into the irrigation canals.

This generous and reliable endowment of surface irrigation water coupled with favorable biophysical conditions supported crop production year-round in the Yaqui Valley (fig. 2.2). Winters are cool, but frosts are uncommon. In theory, it is possible to harvest three crops in two calendar years. Wheat has always been the principle crop grown in the fall–winter growing season (November–April), but it can also be planted during the spring cycle (Finlay 1968). Secondary crops have historically included sorghum, spring cotton, fall maize, and summer soybean. Introduced in the valley in the 1960s, soybeans provided an ideal complement to wheat since they paired a nitrogen-fixing legume (soybeans) with a nitrogen-using grass (wheat).

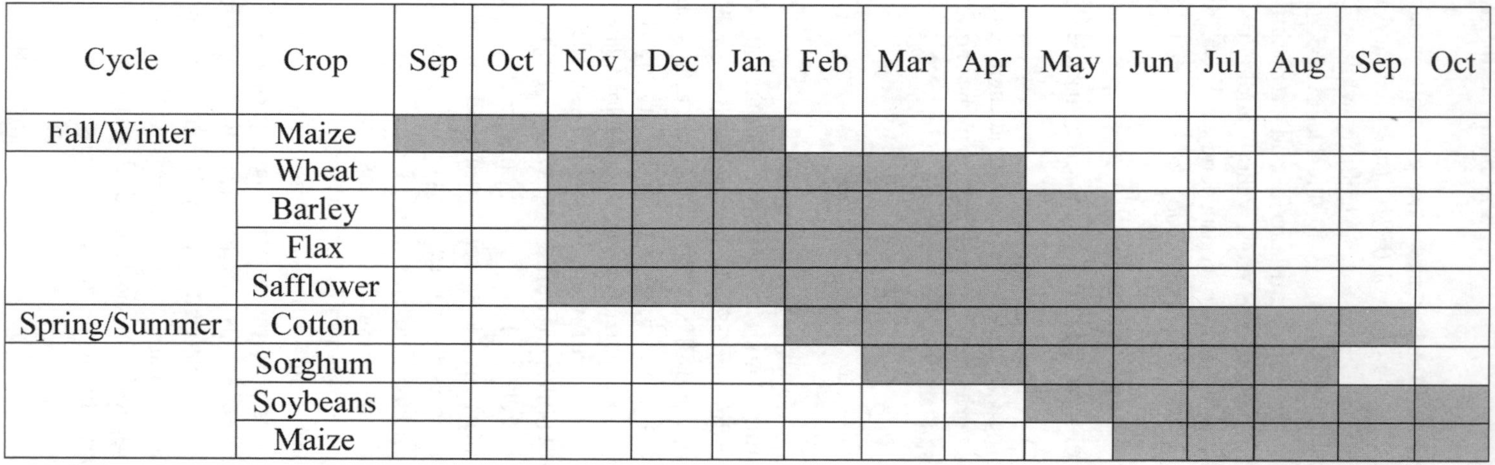

Cycle	Crop	Sep	Oct	Nov	Dec	Jan	Feb	Mar	Apr	May	Jun	Jul	Aug	Sep	Oct
Fall/Winter	Maize														
	Wheat														
	Barley														
	Flax														
	Safflower														
Spring/Summer	Cotton														
	Sorghum														
	Soybeans														
	Maize														

FIGURE 2.2 Agricultural calendar for major crops in the Yaqui Valley (Naylor et al. 2001; adapted from Meisner et al. 1992).

By the middle of the century, the Yaqui Valley had become the most extensive and best irrigated agricultural area in Mexico. It soon became identified as a potential area to conduct international research on wheat agriculture. Working for the Mexican government and the Rockefeller Foundation, renowned agricultural scientist Norman Borlaug, along with an international team of scientists, launched an innovative wheat research program in the Yaqui Valley in 1943. The program's experimental site, Center for Agricultural Research in the Northwest, Mexico, currently known as CIANO, belonged to the National Institute of Agricultural Research (INIA) of the Mexican government and, by 1975, covered approximately 100 hectares of cereal nurseries and trial plots. The main responsibility was to develop new, high-yielding crop varieties through genetic improvements in plant architecture and pathogen resistance, and to promote high-input technologies such as fertilization and pesticides. At that time, farmers were growing tall, introduced wheats such as Mentana (Fischer and Wall 1976). The team developed shorter, higher-yielding wheat varieties that put more energy into grain production and responded better to fertilizer than older varieties.

In 1955, the semidwarfing genes, Norin 10, were bred into wheat genomes and produced the first short cultivars, Pitic 62 and Penjamo 62 (Fischer and Wall 1976). This crossing enabled cereals to stand upright even if excess nitrogen was absorbed, avoiding the problem with lodging that impacted tall varieties when they received too much nitrogen. By 1970, additional experimentation resulted in varieties that ranged between 80 and 90 cm tall and had the yield potential of around 8 tons/ha under the best experimental conditions (Fischer and Wall 1976).[4] Although average farmer yields were consistently about 50 percent less than the experimental yields, some farmers did show the potential to come within 10 percent of the experimental yields. Additionally, average wheat yields for Yaqui Valley farmers demonstrated the same relative increase to those produced in the experimental trial settings.

Yields from these new, short-statured wheat varieties were optimized by increasing the use of fertilizers and irrigation. Fertilization rates averaged 40 kg/ha in 1960 and by 1966 had increased to around 100 kg/ha. They continued to increase to about 250 kg/ha by 2008. The opening of new lands through irrigation development and availability of agricultural credit from both public and private sources contributed to the increase in area under cultivation. Together, these increases in yield and area earned the Yaqui Valley a reputation as the home of the *green revolution* (Freebarin 1963).

By the 1960s, virtually all wheat germplasm planted in the valley descended from the green revolution varieties and continue to be used universally among wheat farmers. In 1966, the wheat research program formally became known as the International Maize and Wheat Improvement Center (CIMMYT), which remains active in the region today.

Mexico's success inspired project researchers, supported by the Ford Foundation, to become dedicated and effective advocates for the Mexican innovation model in other countries; in particular, South Asia, where widespread malnutrition and starvation threatened millions of people. In 1966, having survived one poor harvest but facing another, India took the extraordinary step of importing 18,000 tons of wheat seed from Mexico (CIMMYT 2006). Between 1967 and 1968, wheat harvest increased from 11.3 million tons to 16.5 million tons. Pakistan also began importing the semidwarf wheat seed and adapted high-yield technologies such as pesticides, irrigation projects, and synthetic nitrogen fertilizer. These two countries doubled wheat production between 1966 and 1971.

The resulting social and economic achievements of the green revolution earned Norman Borlaug the 1970 Nobel Peace Prize. The following year, a small cadre of development organizations, national sponsors, and private foundations organized the Consultative Group on International Agricultural Research (CGIAR) to spread the impact of research to more crops and nations. CIMMYT was one of the first international research centers to be supported through CGIAR. CIMMYT, in addition to CIANO, helped markedly increase the productivity of wheat in the Yaqui Valley—increasing average yields for wheat from roughly 2 t/ha in 1960 to its current yield of close to 6 t/ha (Naylor et al. 2001).

Land Stratification and Redistribution

The second half of the century also saw the emergence of diversification and equity issues surrounding land ownership in the Yaqui Valley. During the 1940s to the early 1970s, land redistributions to *ejidos* were minimal as government support moved away from the *ejido* sector back toward the private sector. In 1952, due to discontent among some members of *ejido* communities, the Agrarian Department and the Ejido Bank permitted *ejidatarios* to decide if they would prefer to continue in collective societies or to have the land divided into parcels with individual operations (Freebarin 1963). As a result, three distinct types of agricultural tenure were formed: small proprietor farms, collective *ejidal* farms, and individual *ejidal* parcels.

By 1960, private landowners (larger than 5 hectares) were enjoying wheat yields 50 percent higher than those on small or *ejido* lands. This suggests larger farms were capable of transitioning more quickly into commercialized agriculture while smaller farmers and *ejidos* were mainly producing for subsistence (Hicks 1967). Additionally, private farmers, and to some extent collective *ejidos*, more intensively utilized their land, generally had access to more resources, and were thus able to venture into more diversified crops. Individual *ejidos*, on the other hand, tended to plant only one crop in a given area over the year—either wheat or cotton (Freebarin 1963). Private landowners also exhibited a larger measure of farming efficiency as a result of management factors and internal economic relationships such as production control, marketing effectiveness, greater access to capital resources, enterprise combinations, and internal scale economies. The predominately self-sufficient rural *ejido* economy, however, struggled to be profitable.

Despite the 20 hectares of irrigated land allocated to each *ejido* member, standard of living was still low. Inefficiencies in power, labor, and machinery contributed to the internal disequilibrium among the *ejido* tenure classes. Members overused the land when farming without technical support and were unable to recover costs when machinery had to be rented (Salinas-Zavala et al. 2006). Institutional factors, such as insufficient access of *ejidos* to agricultural credit, and the establishment of a system of mechanization disadvantageous to small farmers also contributed to the disparity between private and *ejidal* farmers. As a result, almost 30 percent of *ejidal* lands were rented to private farmers even though such activity was illegal during this period of time (Freebarin 1963).

In 1976, in response to continued peasant rebellions demanding land, the Luis Echeverría administration redistributed almost 100,000 hectares to local peasants and formed sixty-two new *ejidos* in the region (Cristiani 1984). The valley was divided among *ejiditarios* (55%), private landowners (41%), and colonists (4%) (Naylor et al. 2001). Of the land distributed to *ejidos*, only about one-third of these lands were irrigated. Despite this land redistribution, *ejidos* still faced technological problems because their plots were too small to use the existing technology (Salinas-Zavala et al. 2006). *Ejido* members lacked enough capital to enlarge their farms and increases in land holdings were legally restricted. This encouraged the joining of more individual *ejidos* into collective *ejidos*. Rapid accumulation of machinery and equipment transformed the collective *ejidos* into more efficient and productive communities. However, they still lagged in productivity behind private property holders (Salinas-Zavala et al. 2006).

The other two-thirds of land redistributed during Echeverría's administration were made up of near coastal lands (previously federal territory) that were seen to have little or no productive use at the time (Dabdoub 1980). Some *ejido* fishing communities who already had a long history of living on the coastal land were granted *ejidal* rights over the land. In the mid-1980s, a second major coastal land distribution was carried out by the Salinas administration as part of the Integral Agrarian Program of Sonora (PAIS). Salinas distributed over 5,000 hectares of nonarable coastal lands in southern Sonora to approximately 2,000 new *ejidatarios* to be developed into four shrimp aquaculture "parks." The program was promoted to address the peasant demand for land and to expand economic alternatives for the rural *ejido* sector (Luers et al. 2006). The subsidized pilot projects consisted of groups of *ejido* plots developed as separate shrimp farms but with shared fundamental infrastructure such as pumps, canals, and electrical facilities. Despite these initial efforts, the industry was slow to grow under the hands of *ejidos* due to limited access to credit and the lack of the resources and technical expertise to develop shrimp farms (Luers et al. 2006). The industry did not begin to grow rapidly until almost a decade later when a series of policy reforms opened the region to private investors (Luers 2004; chap. 8).

By the end of the 1980s, the size of a typical ownership unit in the Yaqui Valley was about 25 hectares for privately owned farms and about 10 hectares for *ejido*-owned land (Puente-González 1999). *Ejidos* in the Yaqui Valley accounted for 56 percent of the total agricultural area and 72 percent of all producers in Mexico (CNA 1998; Puente-González 1999), while 40 percent of the total aquaculture area was controlled by *ejidos* (Luers 2004). Credit was readily available for most farmers and the agricultural sector enjoyed a high level of governmental support and involvement. Price supports for agricultural products and input subsidies on water, credit, and fertilizer accounted for 13 percent of the Mexican Federal budget (Naylor et al. 2001; chap. 8). Approximately 33 percent of gross farm income came from these subsidies, which in turn encouraged an increase in inputs. The government was also the primary investor in the shrimp farming industry. These policies were to quickly change beginning in the 1990s when a series of planned reforms shifted responsibility for agriculture and aquaculture from the government to the private and *ejido* sectors, leading to many changes regarding how and by whom land and water resources were being used.

Biophysical and Policy Shocks: 1990–2004

Tumultuous may be an accurate way to describe life in the Yaqui Valley over the last decade and a half; these years are the topic of discussion in the chapters of this book. Individually and in combination, the chapters tell the history of the valley during a period when policy, economic, and institutional reforms drove changes in the use, management, and ownership of both upland and coastal lands, and when biophysical shocks also drove changes. The following section provides a brief overview of these issues, to set the stage for the more in-depth discussions to follow.

In the 1990s, in an effort to integrate rural Mexico into the global economy, the Mexican government implemented a series of policy reforms aimed at reducing government involvement in the agricultural sector by placing responsibility for success more fully in the hands of individual producers (chap. 8). The suite of trade liberalization policies (e.g., NAFTA) left farmers increasingly more vulnerable to the volatile input and commodity markets. Additionally, changes in the *ejido* land-reform program by way of the amendment to Article 27 led to the transfer of both ownership and management of lands from the *ejido* sector back to the private sector (Lewis 2002; Luers 2004).

A series of biophysical shocks (e.g., pest infestation and eight years of drought) also impacted the agricultural landscape—calling into question the economic and environmental sustainability of the valley. During the early 1990s, wheat production remained the dominant cropping pattern in the valley, covering about 85 percent of cultivated land during the fall–winter crop cycle. In 1991, bread wheat accounted for virtually all the wheat grown in the valley (fig 2.3). Shortly thereafter, the spread of karnal bunt (KB), a fungal disease that attacks the wheat kernel, seriously curtailed the exports of bread wheat and wheat seed. Internationally traded wheat originating from KB-endemic areas is severely regulated, and many countries, including the United States, no longer permit the import of bread wheat from Mexico as a consequence of KB. Not surprisingly, KB caused major changes in the autumn portion of the valley cropping system. By 1996, durum wheat, which displays a greater resistance to KB, accounted for more than 80 percent of the wheat grown in the valley (figure 2.3, Naylor et al. 2001; chap. 9).

To complicate matters, the Yaqui Valley suffered from a severe eight-year drought (1997–2004) that put agriculture on the brink of economic disaster (chap. 11). For the most part, farmers in the valley have enjoyed

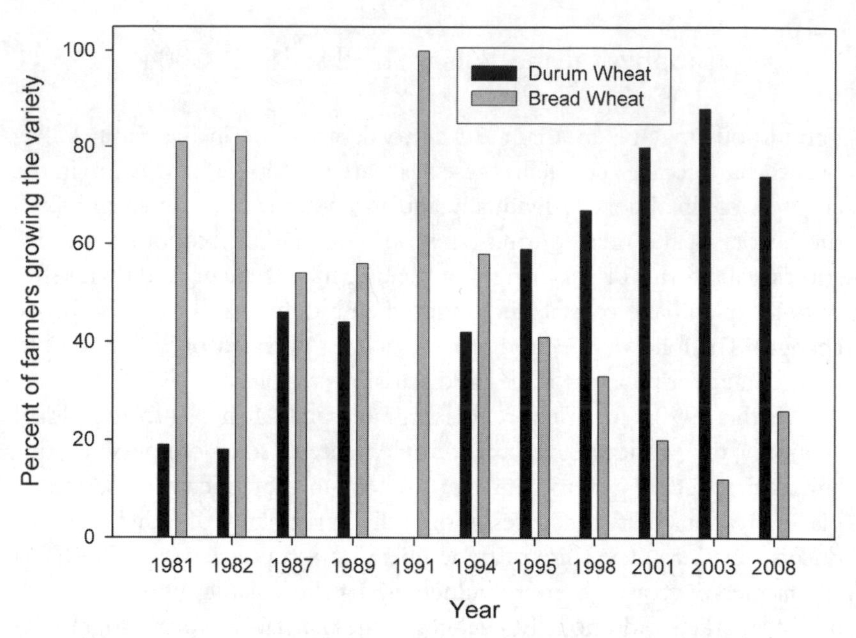

FIGURE 2.3 Wheat varieties grown in the Yaqui Valley, 1981–2008 (Stanford/CIMMYT surveys).

a fairly reliable and stable source of water (about 2 billion m^3 of water annually) provided by the valley's three dams. Beginning in 1994, however, reservoir levels began a mostly downward trend. Between 1995 and 2003, precipitation levels were below average in eight of nine consecutive years, and farmers diverted more irrigation water than they received as reservoir inflow in six of the earlier ten years (Addams 2004). By the height of the drought during the 2003–2004 crop year, no water was left in the Oviachic Dam (fig. 2.4). Only the Yaqui Amerindians, who by presidential degree have the highest priority of access to water from the Yaqui River system, received reservoir water for their territorial lands. The irrigation district had to rely on the pumping capacity of the valley's 180 publicly owned wells, which reduced production to less than 50,000 hectares (chap. 11).

FIGURE 2.4 (opposite page) Annual changes in total water storages in the three Yaqui River reservoirs (MCM is million cubic meters). The two photos are of Oviachic, the downstream reservoir. The first photo was taken in 1992 before onset of the drought, and the second photo was taken at the start of the 2003–04 growing season, toward the end of the drought. (Photos courtesy of Ivan Ortiz-Monasterio.)

1992

2003

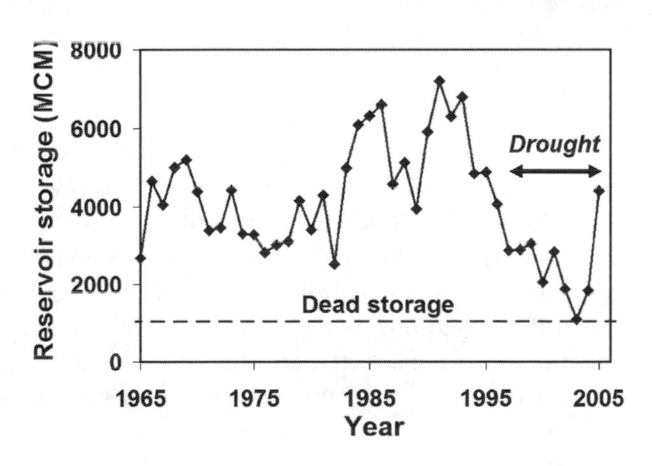
Reservoir storage (MCM)
8000
6000
4000
2000
0
Drought
Dead storage
1965
1975
1985
1995
2005
Year

The drought revealed many of the social inequities that were already plaguing the valley. Water prices rose and only those with extra financial resources could take advantage of water pumped from the region's 400 public and private wells. The irrigation district elected to allocate remaining water reserves based on water demands of the crop—with low water demand and high-value crops given priority—instead of making one valley-wide blanket allocation (McCullough 2005). Farmers who were forced to leave their land fallow received support through an agreement among the Ministry of Agriculture, Livestock, Rural Development, Fisheries, and Food (SAGARPA), the Trust Fund for Shared Risk (FIRCO), and farmer organizations, which paid farmers 905 pesos/ha (about US$80/ha) of land that could not be sown (King 2006). However, by the end of March 2004, 25 percent of the drought relief checks for the fall–winter cycle had not been sent, and government support for harrowing fallow land to prevent weeds was limited (King 2006).

In response to these dire conditions and uncommonly early winter rains, area sown to safflower, a predominately rain-fed crop, increased sixfold during this one year cycle to a total of 61,137 hectares (King 2006). While some safflower was sold to a local oil producer, an export market was not readily accessible. Switching to less water intensive crops may have seemed sensible given the water shortage, but such changes were highly vulnerable to global markets and the skewed subsidy/price support environment that farmers found themselves in, making the transition difficult (chap. 8). During this period, valley farmers, who sought to diversify into various crops and the growing livestock and aquaculture markets, were faced with a range of economic and social constraints including knowledge (chaps. 5, 8, and 9).

The Mid-decade "Snapshot" of the Valley

What did the Yaqui Valley look like at the end of this period? By the end of 2004, roughly 5,000 producers farmed 10,000 individual fields in the valley, with agriculture accounting for 6.6 percent of the Sonoran economy (Pavlakovich-Kochi and King 2006). Only 206,000 hectares of the 233,000 hectares of land comprised within the irrigation district were being planted as some area had been lost to urban growth. Although agricultural management practices had changed during the prior three decades (chaps. 3 and 9), wheat remained the dominant crop accounting for 175,000 hectares of cropland. Summer maize and sorghum were no longer planted, and little land had been allocated to cotton and safflower, 2,100 hectares and 13,700 hectares, respectively (Ortiz-Monasterio, personal communication). The planting of genetically modified organisms (GMOs) was still not permitted even though Mexico could import and consume them.

Fields in the region averaged roughly 20 hectares in size and operational units greater than 50 hectares had become common. The valley had come to resemble the irrigated agriculture of the southwestern United States much more than it does the subsistence-oriented wheat and maize systems of the rest of central and southern Mexico (Naylor et al. 2001). In 2005 *ejiditarios* owned about 50 percent of the valley's land, but actively cultivated less than 20 percent of it. According to a conversation held with AOASS (Association of Producer Organizations in Southern Sonora) director Luiz Signoret, as many as 95 percent of *ejidos* had left farming operations and were renting their land. As discussed in several chapters of this book, this is likely the result of policy reforms and macroeconomic changes in the 1990s that encouraged *ejiditarios* to rent their land to private landholders.

Sustainability transitions in the valley are hard to predict. It is clear that agricultural land-use and management decisions are highly related to the availability of input subsidies, credit, and water supply, and that these are vulnerable to policy, institutional, and climatic changes. A combination of climatic, economic, and biological concerns are now joined by some related to public health. The chapters that follow fill out this short history of the Yaqui Valley to the middle of the first decade of the twenty-first century and provide hints for the way its story will continue to unravel well into the future.

Acknowledgments

I would like to thank the entire Yaqui Project team for giving me a deep understanding of the Yaqui Valley and the complex issues and interactions at play in this dynamic place. A particular thanks to the leadership team of Pamela Matson, Walter Falcon, Roz Naylor, and Ivan Ortiz-Monasterio for seeing this project through in its entirety, and to the Packard Foundation for supporting this work.

PART II

Interdisciplinary Perspectives on Sustainability

Chapter 3

Looking for Win-Wins in Intensive Agriculture

Pamela Matson, Rosamond Naylor, and Ivan Ortiz-Monasterio

Agricultural intensification through the use of high-yielding crop varieties, chemical fertilizers and pesticides, irrigation, and mechanization—known as the *green revolution*— has been responsible for dramatic increases in grain production in developing countries over the past four decades. Expansion of food production into previously nonagricultural lands has likewise increased, but was responsible for only around 10 percent of the overall increased production in the three decades following the beginning of the green revolution (table 3.1; Naylor 1996). Most analysts estimate that close to a doubling of food production will be required in the coming several decades to meet the needs of a growing population and improve food quality and quantity for all, and these increases will most likely take place through continued intensification (FAO 2006a; Bruinsma 2009).

While intensification has played the critical role in increasing food production at a rate that has kept pace with human population growth, and at some level has replaced the need for further land modification (see chap. 4), it carries with it serious implications for off-site environments. Water diversions for irrigation alter hydrology and result in ecological and socioeconomic changes in areas surrounding intensive agricultural systems. The use of fertilizers affects regional and global atmospheric and aquatic chemistry and transport, and thus the ecological status and dynamics of downwind and downstream ecosystems as well as air and water quality and human

TABLE 3.1 Gains in cereal production in the three decades following the introduction of the green revolution, 1961–90.

			Attribution (%)	
	1990 Production (10^6 MT)	*Total increase (%)*	*Increased area*	*Increased yield*
Developing countries	1315	118	8	92
High-income countries	543	67	2	98

Source: Food and Agriculture Organization of the United Nations (FAO) Production Yearbooks, various years.

health. Influx of laborers to agricultural regions and the displacement of previous landowners can drive land-use change in surrounding regions. Despite its benefits in terms of food production, green revolution–style intensification of agricultural systems can have important negative consequences for people, ecosystems, and the global environment (see chap. 4 for more discussion).

In the coming decades, agricultural development will have to proceed with a dual goal—that is, to increase food production while reducing or eliminating negative environmental and human health consequences. This sustainability goal, referred to by Gordon Conway as the "doubly green revolution," requires the development of new agricultural management approaches that incorporate economic, agronomic, and environmental concerns into the decision-making process (Conway 1997; and see the discussion of sustainable agriculture in chap. 1).

The Yaqui Valley provided a wonderful case study for the harmonization of economic, environmental, and agronomic challenges of sustainable agriculture. This region is one of Mexico's major breadbaskets, so agricultural production and its environmental consequences are regionally important. In addition, as the home of the green revolution for wheat, the pattern of management (including increasing rates of fertilizer application) in the Yaqui Valley provided a gauge of what is likely to occur (and indeed, is now occurring) in other high-productivity, irrigated cereal systems of the developing world (Vitousek et al. 2009). While the region has a number of grand sustainability challenges that are addressed throughout this book, we chose first to evaluate the issues related to the use of nitrogen fertilizer.

Globally, application of fertilizer N has increased rapidly in the last several decades, from 9.5 Tg N (9.5 million metric tons) in 1960 to around 80 Tg in 1990, to over 100 Tg today, and is expected to continue to increase, with two-thirds of the application in developing countries by 2050 (Robertson and Vitousek 2009). This anthropogenic N source has already approximately doubled the N inputs to terrestrial systems via natural N fixation (Vitousek and Matson 1993; Vitousek et al. 1997; Galloway et al. 2008), representing an enormous global change. Unfortunately, on average less than 50 percent of the fertilizer N that is applied in any given field is used by the crop, and the rest can be lost to the environment through a variety of mechanisms. Increased losses of nitrate from soils to freshwater and marine systems is the prime driver of eutrophication in many coastal regions of the world and is also a source of poor drinking water quality. Fertilized agriculture is the single most important anthropogenic source of N_2O, accounting for about 60 percent of the anthropogenic sources of this accumulating greenhouse gas (IPCC 2007a). Likewise, fertilization results in elevated emissions of NO, a chemically-reactive gas that regulates tropospheric ozone production and is a precursor to acid precipitation (fig. 3.1; Matson et al. 1997). Several international research programs, including the

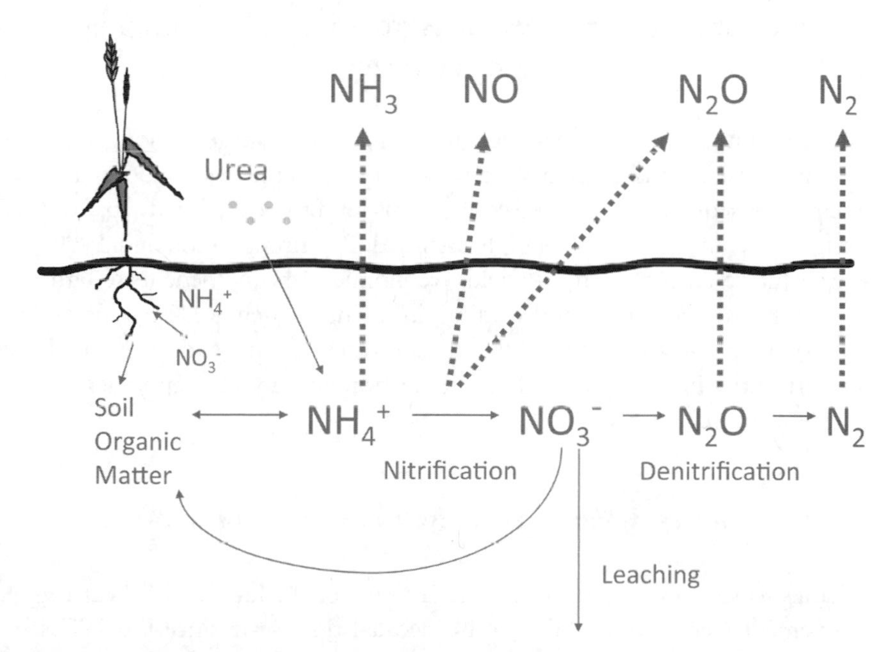

FIGURE 3.1 Simplified version of the terrestrial N cycle depicting gaseous losses following fertilization.

International Geosphere Biosphere Program, the International Nitrogen Initiative, and the Millennium Ecosystem Assessment have identified fertilizer nitrogen losses as key drivers of global environmental change, and key targets for research and action.

Research has shown that management practices can be used to control losses of N; nonetheless, there are many regions within the developed and developing regions of the world where such practices are rarely used (for examples, see Vitousek et al. 2009). While it might be assumed that reducing fertilizer losses should pay off for farmers, assessments of management alternatives in terms of their ability to reduce N losses and yet to be attractive both agronomically and economically were wholly lacking at the time of our study. Our first results suggested that win-wins are possible, but subsequent research illustrated just how challenging it is to make win-win options work in farmers' fields. In the following sections, we discuss the integrated analysis of agronomic, economic, and environmental dimensions of fertilizer application, and then describe several research activities that built upon that work to identify and implement fertilizer management practices that make sense for farmers and the environment.

Integrating Environmental, Agronomic, and Economic Aspects of Fertilizer Use

Most agronomic researchers recognize the close linkages among agronomic decisions and their potential agronomic, economic, and environmental consequences, yet these connections tend to be analyzed separately or in dual combinations, or simply assumed. Beginning in the mid-1990s, we evaluated crop growth and yields, economic costs and benefits to farmers, and nitrogen inputs and outputs in a single integrated analysis and in a single place, and analyzed those data to test the question of win-win opportunities in the Yaqui Valley. This section describes the approach and outcome of that analysis.

Surveying Farmers and Estimating Farm Budgets

Thanks to surveys carried out by the International Maize and Wheat Improvement Center (CIMMYT) on a decadal time scale since the 1950s, we had a reasonable understanding of how typical farmer fertilizer practices were changing in the valley. Our surveys in 1994–95 and 1995–96 indicated that the typical fertilizer practice applied on average 250 kg/ha per six-month wheat crop, with around 180 kg/ha of the N applied approximately a month before planting as urea, and the rest typically applied as anhydrous ammonia in flood irrigation water later in the growing cycle. Application rates had been steadily increasing since the 1950s; while wheat yields had increased only marginally between 1980 and 2000, average fertilizer rates increased by over a third (fig. 3.2). (Some of the reasons for these changes are discussed in chaps. 8 and 9).

Our 1994–95 survey, carried out by Dagoberto Flores, a long-time survey researcher in the valley who was employed by CIMMYT, suggested that farmers had a number of rational reasons for their fertilizer management approach. Many farmers cited the importance of getting fertilizer on early in the crop cycle, both because of the perceived need for nitrogen early in the crop cycle, and to reduce the risk that rains would make the fields inaccessible to machinery and thus delay fertilization until too late to make a difference in yield. They also cited the need to spread labor by completing the fertilizer application before the men and machinery were needed for planting. While some farmers had developed machinery to apply both seeds and fertilizer, it was not a common practice.

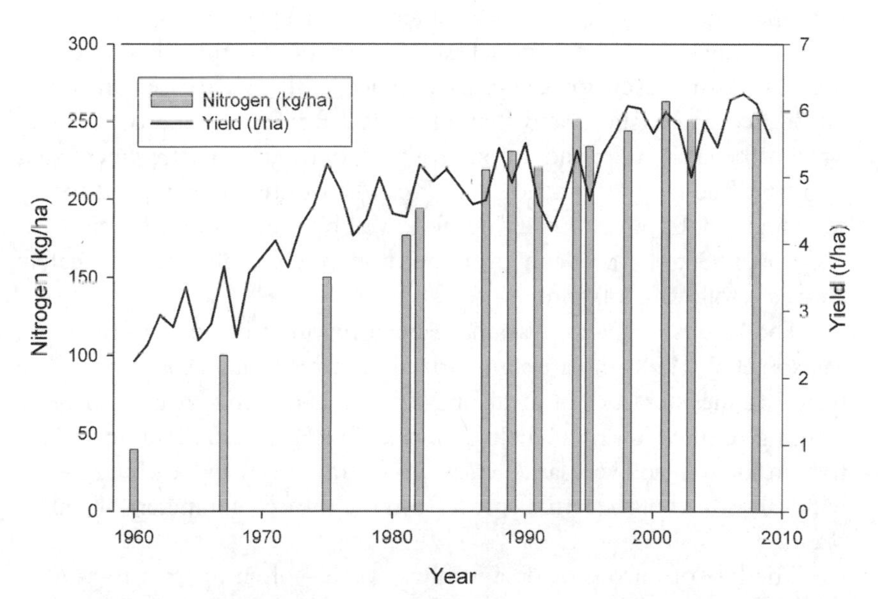

FIGURE 3.2 Average nitrogen application and wheat yields in the Yaqui Valley, 1960–2008 (Stanford/CIMMYT surveys).

As in many high-productivity agricultural systems of the developing world, the dissemination of green revolution technologies initially provided farmers in the Yaqui Valley with modern seed varieties and highly subsidized nitrogen fertilizers; earlier surveys of farm budgets suggested fertilizer was not a significant cost. In the several years prior to our 1994–95 survey, however, the reduction of subsidies (in real terms) had been dramatic (see chap. 8 for a discussion of the policy changes leading to this reduction). Our economic analysis of farmers' costs and returns for both 1994–95 and 1995–96 wheat seasons indicated that fertilization had become the highest direct production cost in the Yaqui Valley farm budgets (table 3.2). During the two years of this analysis, fertilization exceeded even the costs of land preparation, which traditionally have represented the largest cost category in this highly mechanized system; just five years prior, the cost of land preparation was 50 percent higher than that of fertilization.

Studying the Effects of Fertilizer Practices

In the 1994–95 and 1995–96 wheat cycles, we evaluated yields, grain quality, soil nutrients, gas fluxes, and solution losses of N prior to and following fertilizer additions in experimental plots at the CIMMYT field station and in farmers' fields. We included several fertilizer management approaches in our experiment: the conventional farmers' practice for the valley, as determined by farm survey; three alternative practices that were based on agronomist recommendations and that added less fertilizer N and/or fertilizer later in the crop cycle; and a nonfertilized control. In our treatment that simulated the farmers' practice, 187 kg N/ha of urea were applied to dry soils one month prior to planting, followed by preplanting irrigation; an additional 63 kg N/ha of anhydrous ammonia were applied approximately six weeks following planting.

The results of the analyses have been reported in other forms (e.g., Matson et al. 1998) and are summarized in some detail in chapter 10. The typical farmer's practice of fertilizing dry soil and then irrigating prior to planting resulted in extremely high levels of inorganic ammonium and nitrate in the soil, and very large emissions of trace gases and leaching losses, with 40 percent of the fertilizer lost from the top 15 centimeters of soil by planting.

The loss of nitrous oxide and nitric oxide—often referred to as *trace* gases, because they are present in the atmosphere at parts per billion levels and their fluxes are likewise typically very small—told an important story.

TABLE 3.2 Costs and returns for the 1994–95 and 1995–96 wheat cycles. Data collected during on-farm surveys.

*Factors (N$/ha)**	*1994–95*	*1995–96*
Costs		
Land preparation	342	345
Planting	248	369
Fertilization	519	1,041
Irrigation	96	300
Insect and weed control	177	238
Harvest	156	454
Other costs [1]	1,710	1,560
Total	3,248	4,307
Returns		
Gross revenue from yield (pre-income tax)	4,294	8,450
Profit		
Returns to management (after income tax)	1,046	3,413

Source: Matson et al. 1998. ©2011 by AAAS.

*Costs, returns, and profits are shown in new pesos (N$) per hectare. The average exchange rates were N$3.2/US$ in 1994, N$5.58/US$ in 1995, N$7.55/US$ in 1996.

[1] Other costs include interest on credit, crop insurance, salary of the field manager, independent technical assistance, producer's organization fees, land rental, taxes, and subsidies.

The farmers' practice resulted in very large emissions of both gases in both years (fig. 3.3), with preplanting gas fluxes summing to 5.6 and 4.6 kg N/ha in the 1994–95 and 1995–96 wheat cycles, respectively, representing over 40 percent of total trace gas losses through the crop cycle. Measured gas emissions were among the highest ever reported. Measured leaching losses were likewise high. Our analyses suggest that the typical farmer practice leads to a loss of 14 to 26 percent of the applied fertilizer from the top meter of soil in the month prior to planting (Riley et al. 2001); some of that nitrogen was likely stored in deep soils, and some was lost from deep soils to the groundwater and surface water systems.

The results of the alternative fertilization treatments also told an interesting story. The alternative practice that applied 250 kg/ha N, with 33 percent before planting, 0 percent at planting, and 67 percent after planting,

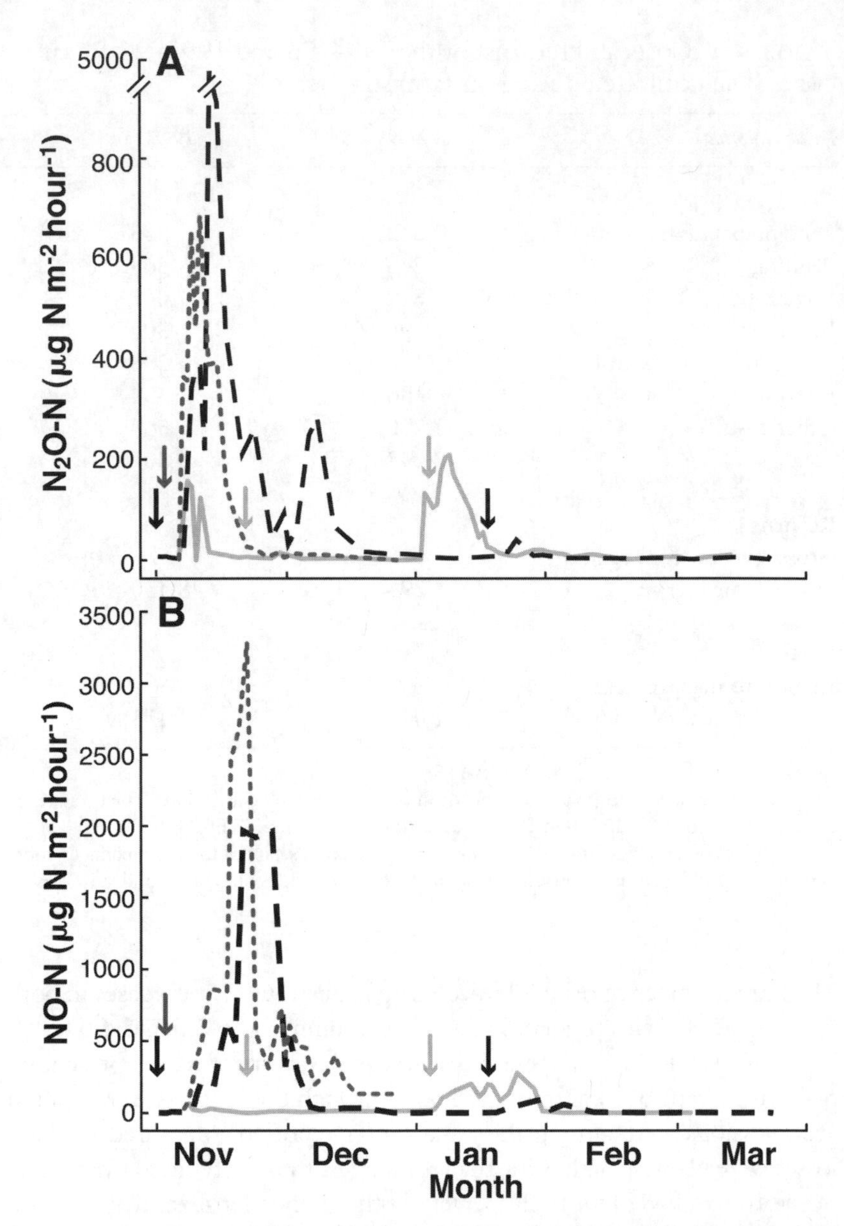

FIGURE 3.3 Changes in the emissions of N_2O and NO (µg N m^{-2} hour^{-1}) from the soil surface in the farmers' practice over the 1994–95 (black dashed line) and 1995–96 (dark grey dashed line) wheat cycles, and for the best alternative (light grey solid line) in the 1995–96 wheat cycle. (A) N_2O-N emissions. (B) NO-N emissions. Fertilizer applications in the different treatments and years are indicated by arrows (coded to match the flux data) (Matson et al. 1998). © 2011 by AAAS.

had similarly high losses as the farmers' practice. Losses in the early season seemed to be critical. The alternative practice that added 250 kg/ha N, with 33 percent at planting and 67 percent postplanting, lost at least 50 percent less N gas than the farmers' practice. The "best" alternative with respect to reduced N_2O and NO emissions applied a total of 180 kg N/ha, with 33 percent at planting and 67 percent six weeks postplanting (fig. 3.3). In this treatment, total fluxes of N_2O and NO, summed over the 1995–96 wheat cycle, were 0.74 kg N/ha—an order of magnitude less than that occurring with the farmers' practice. Modeled nitrate losses were also 60–95 percent lower (see chap. 10).

The interplay between the timing of fertilization and irrigation was critical to nitrogen losses in this site. When fertilizer was added to dry soils, only very small changes in inorganic N concentrations or N gas emissions were measured. With preplanting irrigation—a practice that only became common in the 1990s and was highly effective as a means of weed control—however, the fertilizer was rapidly converted from urea to ammonium followed by the oxidation of ammonium to nitrate by bacteria, a process that is associated with losses of nitrous oxide and nitric oxide gas emissions as well as nitrate leaching losses. By the time planting occurred, trace gas fluxes had returned to low levels, and only low levels of inorganic nitrogen remained in the top layers of soil. In other words, lots of the action in terms of fertilizer losses occurred before the seeds were planted, and those losses were very important for their environmental consequences as well as in terms of lost investment. The timing issue sets up a potential conundrum for farmers: to dramatically reduce fertilizer losses and environmental consequences, they would have to develop approaches that would avoid preplanting fertilization. We'll return to this issue later.

Emissions of N_2O and NO under the farmers' practice are large relative to those observed in many other studies. However, they represent just two of several important pathways by which nitrogen can be lost from terrestrial ecosystems; others include ammonia volatilization and dinitrogen gas flux, both of which, according to our rough measurements and estimates, were very high after preplanting fertilization and irrigation. Ammonium losses in tail water were also high, but they were likely to occur later in the crop cycle during periods when ammonia was added to irrigation water. Our further studies evaluated these and other loss pathways and their environmental consequences (see chap. 10 for more of the story), and found, for example, a continuing cascade of nitrogen effects through the drainage streams to the sea, where nitrogen carried in irrigation water led to vast phytoplankton blooms.

To farmers, total loss of nitrogen is perhaps of more interest than specific trace gas or solution losses, because the amount of wasted fertilizer represents a financial cost. In order to determine quantitatively the total loss of fertilizer N, we applied ^{15}N-labeled urea (in place of the fertilizer) in isolated plots that were otherwise treated like the experimental plots; at the end of the crop cycle, the isolated plots were harvested and ^{15}N recovery in soil and plant components was measured. In the experimental farmers' practice and in our best alternative, proportional recovery of the applied N in plants was 46 percent and 57 percent, respectively. In later studies, discussed in chapter 8, we found that even with equal amounts (225 kg/ha) of fertilizer applied with both practices, preplanting fertilization led to a 41 percent nitrogen use efficiency (defined as ratio of added N recovered in plants) while fertilization at planting was 50 percent. Moreover, in farmers' fields using the conventional practice, recovery of nitrogen from fertilizer was only 31 percent. The elimination of preplanting fertilization and application of most of it later in the crop cycle seemed to make financial sense.

Is There a Win-Win Situation Here?

Fertilizer use and loss are just one component of farm budgets, and farmers typically focus not only on costs but on the balance between costs and expected income under some degree of price and production uncertainty. For wheat farmers in the Yaqui Valley, yield of good quality wheat provides the essential income. Yields reported in our socioeconomic surveys in 1994–95 and 1995–96 ranged from 3.1 to 7.3 t/ha, with average values of 4.9 and 5.3 t/ha for the two seasons, respectively. Mean yields in our simulated farmer practice were 6.08 ± 0.18 and 6.07 ± 0.28 t/ha in 1994–95 and 1995–96, respectively. Our best alternative, which added 180 kg N/ha in contrast to 250 kg N/ha in the farmers' practice, resulted in yields that were not significantly different (6.16 ±1.3 t/ha). Likewise, grain quality (estimated as the protein concentration in grain) in the alternative was not significantly different from the farmers' practice (14.87% versus 14.83%, respectively).

Given the importance of fertilizer in the valley farm budgets, we evaluated the extent to which increased fertilizer efficiency represented a significant budgetary savings to the farmers (Matson et al. 1998). In contrasting the farmers' practice with our best alternative, we found that the alternative resulted in a savings of M$414–571/ha, or US$55–76/ha at then existing prices and exchange rates. These values, which resulted from lower fertilizer applications and reduced loss of fertilizer, were equivalent to 12–17 percent savings of after-tax profits from wheat farming in the valley. Given these good results, this best practice was tested in twenty-eight on-farm trials over the following two crop cycles, the average results of which supported the conclusion that 180 kg/ha (applied at planting and one time after) was more profitable than 250 kg/ha (applied primarily preplanting). Altogether, our results suggested that alternative fertilizer practices could reduce trace gas and total losses of fertilizer and maintain yields. They also suggested that the best alternative, which matched crop demand and fertilizer application, ultimately could save money and perhaps might allow Yaqui Valley farmers to remain competitive in an era of economic liberalization and expanding free trade.

At the time that we completed and published this first study (and reported our results in farmers' workshops in the Yaqui Valley), we wondered if the potential cost savings might over time induce a shift in technology and management toward fertilization later in the wheat cycle. We recognized, however, that many things could get in the way of implementation of the new practice, and that farmers could perceive risks to doing so. While we would have liked to declare victory and move on, we instead set out to track farmers' research, and then revise our plans as necessary.

The Challenge of Making Win-Wins Real

Our next surveys in 1998 and 2001 indicated that, if anything, farmers were applying more N, not less, despite continued increases in fertilizer prices. In 2001, the average application rate was 263 kg/ha. At this point, our research pursued several new directions in an attempt to understand the barriers to adoption of the new practice or any more environmentally friendly practice.

First, as is discussed in more detail in chapter 9, we delved into the farmers' perceptions about physical impediments to applying the first fertilizer application at or after planting. A very small percentage of farmers—around 10 percent—reported applying their first fertilizer at planting. Generally, as our early survey indicated, the concerns of the other 90 percent reflected perceived and real risks about the interaction of rain and fertilization. For example, carrying out fertilization at planting time extends the planting time because moist soils (from the preplanting irrigation) slow the operation of machinery compared to application on dry soil. Likewise, delaying the application of more of the fertilizer to the postplanting period, when the likelihood of rain is significantly greater, increases the risk of insufficient nitrogen for the crop just at the time when it is needed most. However, the fact that 10 percent of farmers were no longer doing preplanting fertilization suggested that the risks were not viewed similarly by all farmers. Our conversations with farmers suggest that the farmers managing smaller acreages were more likely to switch to no preplanting fertilizer. Farmers operating on larger acreages (which often consist of fragmented pieces of land located across the valley) did not, probably due to the logistical challenges of management.

Other research by team members uncovered the fact that interannual variability in growing season temperatures and spatial variability in residual soil nitrogen amounts and availability, among other factors, cause significant uncertainty for farmers (chap. 9). These analyses concluded that, over longer time scales, it can be economically sensible for farmers to overapply nitrogen fertilizer, assuming no regulations or costs associated with the environmental negatives (Lobell et al. 2004).

Finally, and somewhat to our dismay, we discovered that the farmer credit unions—farmer associations that provide members with credit, seeds, fertilizer, other inputs, insurance, postharvest storage, marketing, and technical assistance—to which most farmers belong, were encouraging the use of increasing amounts of fertilizer (see chap. 5 for a detailed discussion about the role of the credit unions). In fact, credit unions regulate the farm management activities of their members by providing crop insurance that is contingent on compliance with official production recommendations. These recommendations are apparently formed on the basis of official technical packets from the local agricultural research stations as well as the experiences of the union's technicians and their members. Their recommendations may have reflected the previously mentioned concerns about risks and variability, thus encouraging all farmers to manage for the optimum year and soil, but there is certainly also a financial incentive for unions to encourage high levels of fertilizer use (and sales) in the valley.

Given the real concerns of farmers and credit unions about temporal and spatial variability, we began to direct our efforts toward the development of diagnostic tools that would assist farmers in decision making under uncertainty. The story of the development of the "GreenSeeker" sensor (fig. 3.4) for site-specific nitrogen management is told in detail in chapter 9. In brief, the instrument measures reflectance in several wavelengths that, in combination with crop growth patterns and comparisons with crops growing in narrow fertilized strips within the field, allow the estimation of yield potential and nitrogen demand of a crop beginning forty days after

FIGURE 3.4 The GreenSeeker. (Photo courtesy of Ivan Ortiz-Monasterio.)

planting during the cropping period. Thus, farmers have a tool that would allow them to determine when and how much fertilizer is needed in specific fields under specific climate conditions.

But as we've seen before, the presence of a tool or technology does not imply use. To encourage broader use of the GreenSeeker, we knew we would have to engage the credit lenders who hold many management strings, as well as the farmers who ultimately carry out the practice. As in many other aspects of this work, the relationships that our team member Ivan Ortiz-Monasterio had with the local community, and the trust they placed in him, really paid off. We engaged six farmer unions in analysis and trials of the system, each paying 50 percent of the costs of an instrument that could be dedicated to their members. Six years of evaluation in farmers' fields showed average reduction of 70 kg N/ha without a yield reduction. During crop cycle 2008–09, this technology was being used on approximately 7,000 hectares. Ultimately, site-specific management practices that tools like the GreenSeeker allow are one part of the path to sustainability. They don't solve all the problems, but they move individual farmers a step or more in the right direction.

There's still a long way to go. The recently enormous price volatility (for both wheat and inputs) makes it hard for farmers to engage new technologies and feel the benefits of preventing losses of fertilizer or other inputs. Likewise, climate interactions with water management and cropping decisions loom large on the horizon for the Yaqui Valley, as do possible changes in national subsidies and new institutional reforms. Ultimately, change in policies, markets, and climate, among others, will induce change in farmer decisions about crops and management, and are also likely to result in changes in terms of environmental concerns. For example, if the current tendency for corporate contracts with farmers to produce bread wheat continues, preplanting fertilization will likely be dropped (because postplanting fertilization is associated with improvements in grain quality), followed by a likely dramatic reduction in nitrate runoff and nitrogen gas emissions. Likewise, changes in crop irrigation management in response to climate changes could carry with it possible reductions in nitrogen losses. Increasing awareness of the off-site consequences of agriculture (for example, in terms of air pollution, euthrophication in the ocean, negative effects on drinking water quality, and negative consequences for human health) could push agricultural management in new directions. Over time, under policy and environmental pressures, farmers could diversify or completely change their systems to new crops and new management regimes in ways that could benefit both yields and environment, but only if knowledge is available to help them do so (see chap. 5). These interactions in the Yaqui Valley are illustrative of the interactions between environmental and development goals being played out around the world, especially in rapidly changing places like China, Brazil, Indonesia, Pakistan, Bangladesh, other regions of Mexico, and many other countries in which our team members are engaged (Vitousek et al. 2009; Chen et al. 2011). There are many forces at work, no one magic solution, and no one all-encompassing, win-win opportunity in agriculture. Perhaps the best we can do is develop and foster interest in new systems and new practices that are agronomically and environmentally sustainable, encourage governing institutions to do the same, and help provide tools and approaches that make sense for farmers and for the environment.

Acknowledgments

We thank W. Falcon, P. Vitousek, D. Winkelmann, R. Fischer, and S. Rajaram for discussions and support. C. Billow, J. Moen, M. Cisneros, and

J. Panek managed and performed the field and laboratory collections and analyses. P. Brooks assisted with gas chromatography. B. Avalos, E. Rice, D. Flores, and J. Harris assisted in the socioeconomic surveys and analysis. D. Saah, J. Perez, S. Zuniga, L. Mendez, B. Ortiz, N. Placencia, H. Farrington, P. Vitousek, M. Mack, K. Lohse, T. Benning, S. Lindblom, and S. Hall assisted in the field and laboratory. We thankfully acknowledge funding from the US Department of Agriculture (USDA) Ecosystems Program, National Aeronautics and Space Administration (NASA) Land Cover/Land Use Change Program, the Andrew Mellon Foundation, and the MacArthur Fellows Program (PM); the Pew Fellows Program of the Pew Charitable Trust, Ford Foundation, and Hewlett Foundation (RN); and CIMMYT, Ford Foundation, and Hewlett Foundation (IO-M).

Chapter 4

Ecosystems and Land-use Change in the Yaqui Valley: Does Agricultural Intensification "Spare Land for Nature"?

PAMELA MATSON AND PETER JEWETT

Humans have had impacts on the land surface as long as they've inhabited the planet. However, in the past half century of rapid population growth and industrialization, the degree, extent, and rate of change has become so great as to be significant at a global scale. One of the most amazing statistics of global environmental change is the estimate that *half* the ice-free land surface has been modified—that is, completely changed—by human activities (Kates et al. 1990). Change in the land surface has wide-ranging impact on the functioning of the planetary system and the provision of goods and services for humanity (MA 2005), and thus the study of land change, including its causes and consequences, has become a central area of research in the global environmental change and sustainability science literature (Turner et al. 2007).

Agriculture has been a dominant driver of land-use change. Historical changes, reconstructed through the use of remote sensing over the past several decades together with hindcasting of vegetation cover and historical population density, illustrate the dramatic increases in cropland and pasture lands over the past three centuries (table 4.1; Lambin et al. 2003), balanced by a decline in steppes, savannas, grasslands, and forests. Over the past forty years, however, the more critical agricultural land-use change has not been land-cover change but land-use intensification. Most of the dramatic increase in the world's agricultural production from 1961–90 was due to increasing yields per area rather than increasing area, thanks

TABLE 4.1 Historical changes in land use/cover at a global scale over the last 300 years.

	Cropland (10^6 *ha*)	*Pasture* (10^6 *ha*)	*Forest/woodland* (10^6 *ha*)	*Steppe/savanna/ grassland* (10^6 *ha*)
1700	300 to 400	400 to 500	5,000 to 6,200	3,200
1990	1,500 to 1,800	3,100 to 3,300	4,300 to 5,300	1,800 to 2,700

Source: Lambin et al. 2003. © 2011 by Annual Review of Environment and Resources.

primarily to dramatic increases in fertilizer use and irrigation (see chap. 3, table 3.1; Naylor 1996; Tillman et al. 2002).

One potential benefit of increasing yields rather than area is that natural areas can then be reserved for the benefit of biological species other than humans, as well as for the goods and services they provide to humans. This potential relationship has sometimes been referred to as "sparing land for nature" (figure 4.1; World Bank 1992; Waggoner 1995). While the argument that increasing yields is necessary in order to save natural lands makes inherent sense, the question remains whether, in fact, it does result in the saving of natural lands. Indeed, there is a wide range of consequences resulting from intensive agriculture practices that could result in just the opposite occurring at local or regional scales (Matson and Vitousek 2006; Rudel et al. 2009).

Some of the potential negative consequences of intensive agriculture for soils, water, and atmospheric systems, both at the agricultural site and off-site, are well known (Matson et al. 1998; Matson and Vitousek 2006; Robertson and Vitousek 2009, and other reviews). Off-site effects are typically linked to the high inputs of fertilizers and water that accompany increasing yields. The use of nitrogen fertilizer has led to a range of problems of groundwater pollution and coastal degradation (Vitousek et al. 1997; Robertson and Vitousek 2009; chaps. 3 and 10). Nutrients that leach or run off from agricultural fields can be transferred to downstream ecosystems where they can influence biogeochemical processes and can lead to changes in biological diversity and negatively affect economically important resources such as fisheries and tourism (NRC 2000; Carpenter and Gunderson 2001; Howarth 2008). Likewise, atmospheric transfers of nutrients or air pollutants can affect the functioning of downwind ecosystems, including agricultural as well as natural ecosystems (fig. 4.2; Chameides et al. 1994; Matson et al. 1997; Holland et al. 2005; MA 2005).

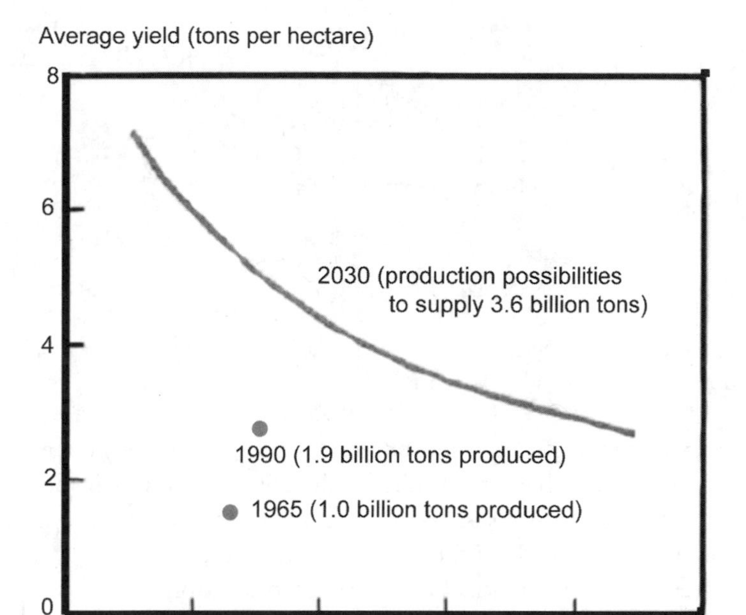

FIGURE 4.1 World production of cereals to feed a growing population: past performance and the future challenge (adapted from World Bank 1992). © 2011 World Bank.

Water withdrawals from surface water systems can interrupt downstream aquatic ecosystem processes by reducing flows, altering salinity and other chemistry, changing the timing of flows, or preventing the movement of animals. Today, nearly 85 percent of annual water withdrawals globally are for agriculture (Foley et al. 2005), and as a result, many rivers now run dry at least part of the year.

An additional dimension of off-site consequences of agricultural intensification and increasing yields has to do with the more indirect relationship of intensification with land-use expansion (extensification into previously unmanaged lands). Agricultural intensification and increasing yields may lead to increased population growth within, or migration to, a region; along with the growth of related industrial and other economic activities, such population increases can put new demands on land in the areas surrounding the intensively managed districts (see Angelsen and Kaimowitz 2001 for numerous case studies illustrating these points). Likewise, change in land management through mechanization and increase in land holding

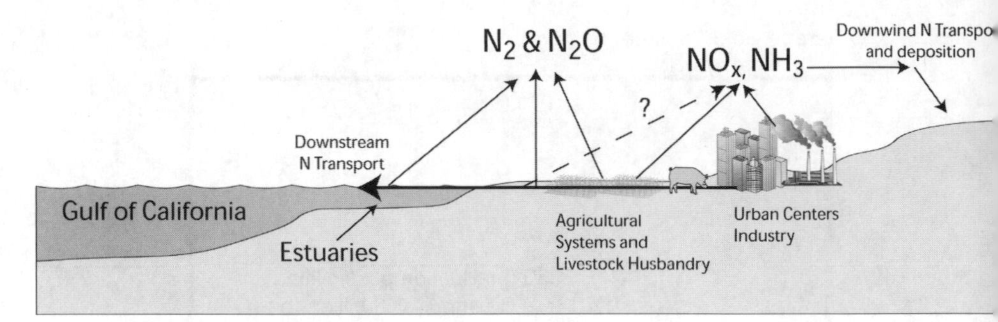

FIGURE 4.2 Atmospheric transfers of nutrients or air pollutants can affect the functioning of downwind ecosystems.

sizes may lead to loss of agricultural jobs and may potentially displace people from lands that they have traditionally worked (Kaimowitz and Smith 2001; Matson and Vitousek 2006). These have the potential to influence land-use changes outside of the intensive agricultural area.

Given both the potential benefits of intensification (in terms of reducing land conversions while producing more food) and the potential negatives due to off-site effects, what are the dynamics occurring in the Yaqui Valley? We examined land-use changes over several decades in an area of rapid agricultural intensification in order to evaluate the specific and empirical rather than just theoretical linkages among intensification, increasing yields, and off-site consequences for nonagricultural ecosystems, and to draw conclusions about the importance of an intensive agricultural system on surrounding natural ecosystems. We used a combination of socioeconomic surveys and interviews, remote sensing analysis, and ecological/biogeochemical measurements to understand the dimensions of change in the past forty years—the agricultural changes occurring, the factors driving them, and the off-site consequences in terms of saving land for nature. Our effort is incomplete—many pieces are missing from a thorough analysis—yet the story is still interesting.

Our attention was paid primarily to the period of *intensification* in the Yaqui Valley, after land had already been converted from natural ecosystems to agriculture. Nonetheless, it makes sense to start the discussion with a picture of what the Yaqui Valley looked like prior to the *green revolution*.

The Landscape Prior to Intensive Agriculture

In 1935, botanist Forrest Shreve described the coastal thornscrub vegetation of the Yaqui Valley:

> South of the Yaqui River there is much of the desert in which it is not possible to walk a straight course for more than a few yards, or to go very far without the use of a machete. The stature of the trees has increased very little, but their number per acre is much greater [than the desert scrub of the Sonoran Desert]. Smaller plants are no longer clustered more densely in the shade of the trees but grow everywhere. Streamways are no longer discernible at a distance by their fringe of heavier vegetation but only by their slightly taller trees and their dense thickets of shrubs and vines. The cacti of the open desert begin to give way to species which endure the shade of the thin tree-tops. The soil reacts to the heavier vegetation and contains much more organic matter. (Shreve 1935)

The Yaqui Valley, bounded to the north by the Yaqui River and to the south by the Mayo River, comprises the southern most extent of the Sonoran Desert. The Sonoran Desert covers 223,000 square kilometers, spanning much of the Mexican state of Sonora, along with parts of the southwestern United States and Baja California, and is home to the greatest diversity of vegetation of any desert in the world (Nabham and Plotkin 1994). The region is characterized by heavy, localized summer rains and widespread, light winter precipitation, while also being subject to long periods of drought.

The Yaqui River is the largest river in Sonora, draining an area of approximately 72,000 square kilometers, including most of the west slope of the Sierra Madre Occidental and portions of Arizona and the Mexican state of Chihuahua. While the river never had year-round flow in the lower section, during flood season the head of the delta was described as rivaling that of the Colorado River (Shreve 1951). Driven by the bimodal annual rains of its huge watershed, the Yaqui River typically flooded twice annually prior to human alteration. The seasonal flooding of the Yaqui River over tens of thousands of years formed the flat, delta plain known as the Yaqui Valley, which extends southeast from the river's active floodplain over an area of 1,800 square kilometers to its eastern boundary near the town of Ciudad Obregón (Shreve 1951; West 1993).

The active floodplain of the Yaqui River, which contained sandy loam soils rich in depositional organic matter, was cultivated for hundreds of years by the Yaqui Indians prior to being rendered inactive by upstream dams (West 1993). Cottonwood and willow forests lined the lower reaches of the Yaqui River and supported a high diversity of birds (Russell and Monson 1998). In the Yaqui Valley to the south, the alluvial soils tended toward less fertile, heavy clays that were dominated by an abundance of spiny vegetation called coastal thornscrub (Dabdoub 1980).

Coastal thornscrub is an intermediary between desertscrub, which inhabits drier regions of the Sonoran Desert to the north, and tropical deciduous forest that is present to the east and south of the Yaqui Valley. Prior to large-scale agriculture, coastal thornscrub formed 100 percent vegetative cover in the Yaqui Valley and was characterized by an abundance of leguminous trees, dense shrubbery, vines, and cacti. Trees such as mesquite (*Proposis* spp.), acacia (*Acacia* spp.), and palo verde (*Cercidium* spp.) reached heights of 6.5 meters and were accompanied by shrubbery, such as ocotillo (*Fouquieria* spp.) and a number of species of aborescent cacti, including the etcho (*Pachycereus pectin-aboriginum*) and pitahaya (*Stenocereus thurberi*) (Gentry 1942; Shreve 1951).

The thornscrub vegetation had many dietary uses for the Yaqui Amerindians, as well as other people in southern Sonora. Mesquite pods, rich in carbohydrates and protein, were ground and used for making bread. Fruit from the various cacti provided vitamins, such as ascorbic acid. The thornscrub also supported white-tailed deer and smaller mammals like rabbits and rodents, which were hunted by the region's inhabitants (West 1993).

Near the Gulf of California, coastal thornscrub gave way to sand dunes, wetlands, bare ground, lagoons, and tidal flats of the coastal zone, which stretched from north of the Yaqui River delta to southern Sonora and beyond. The brackish tidal flats and sandbar islands were colonized by mangrove communities, which exhibited a nearly continuous range along the 250-kilometer stretch of coastal zone prior to the disruption of the seasonal flooding of the Yaqui River (Felger et al. 2001). The coastal zone extended up to 5 kilometers inland from the coast and provided critical habitat for migratory and resident water birds, marine mammals, fish and shellfish populations (Flores-Verdugo et al. 1992). The biological richness of the coastal zone continued into the Gulf of California, designated by UNESCO as a World Heritage Site due to the fact that it is home to 891 species of fish, 39 percent of the world's marine mammal species, and fully one-third of all cetacean species on Earth (chap. 7).

Moving east from the Yaqui Valley into the foothills of the Sierra Madre, coastal thornscrub transitioned to tropical deciduous forest, which is found at the northernmost distribution of its range in the Western Hemisphere. The tropical deciduous forest to the east of the Yaqui Valley primarily occupies streamways and barrancas (cliffs and gorges), characterized by increased moisture and habitat diversity, before becoming more pervasive as it stretches as far south as Paraguay. A rich biodiversity of plant species, such as kapok (*Ceiba acuminate*) and fig species, are found in the tropical deciduous forest, as well as fauna such as parrots, boa constrictors, and mountain lions (Martin and Yetman 2000).

The Agricultural Landscape of the Last Fifty Years

While the Yaqui Valley's floodplain had been farmed nonintensively for hundreds of years (chap. 2), broadscale land clearing and intensification of the region occurred in the mid-twentieth century. With the completion of the Angostura dam in 1942 and the Oviachic dam in 1953, the seasonal flooding of the Yaqui River was contained, allowing for year-round irrigation. By 1955, 210,000 hectares were in cultivation, with the only natural lands remaining outside the irrigation district boundaries—in the coastal zone and in the foothills of the Sierra Madres. We were interested in the ongoing changes in those nonconverted areas, but we were also interested in the dynamics occurring within the agricultural district itself. In the following sections, we first discuss what we learned about the agricultural district land-use dynamics, and then about the off-site ecosystems.

The Agricultural District

Using North American Landscape Characterization satellite imagery for 1973, 1986, and 1994 we evaluated change in area under intensive agriculture and found only a 6.3 percent increase in irrigated area over the thirty-year period. This relatively small change reflects the early green revolution development of this site—the irrigation system of reservoirs, canals, and drainages was mostly in place by the end of the 1950s, and set the stage for the agricultural development that came after. Indeed, total allocation of irrigation water was relatively constant (2640 +/– 950 million cubic meters) during the period of 1970–96 (Addams 2004).

Despite the fact that the area under irrigation has not changed substantially since the beginning of intensive agriculture, the Yaqui Valley has undergone dramatic changes in crop varieties and cropping systems, fertilizer inputs, and yields—the hallmarks of intensification. Wheat yields averaged around 2 tons/ha in 1960; during the time of our remote sensing analysis, it increased to nearly 6 tons/ha (see fig. 3.2 in chap. 3). This increase can be explained in part by a steady influx of modern, higher yielding wheat varieties provided by CIMMYT (International Maize and Wheat Improvement Center) to farmers here and around the world (Naylor et al. 2001; chaps. 7

and 8). Likewise, the dramatic increase in fertilizer inputs (see fig. 3.2 in chap. 3) have contributed to increased yields—agronomists in the valley suggest that the optimal N fertilization rate for the high-yielding varieties of wheat is 167 kg/ha/wheat cycle. Average application rates in the valley had reached that point by 1980, yet input rates continued to increase.

These changes in inputs can be explained at least partially by policies put in place by the Mexican national government. In the 1970s and 80s, a set of domestic policies, that included price supports, input subsidies, and consumer subsidies, adding up to 18 percent of federal budget, supported intensification (Naylor et al. 2001; chap. 8). Based on our analysis, approximately 33 percent of gross farm income came from subsidies in 1980s, and encouraged continued increase in inputs. During the 1990s, these and other subsidies changed with the suite of liberalization policies (discussed in detail in chap. 8).

An additional dramatic change that occurred in the valley during the later part of our study period is the emergence of the livestock sector. During the decade of the 80s, there was an estimated fifteenfold increase in pigs and poultry in the valley. By 1996, production of pigs and poultry was approximately 350,000 head and 6 million birds, respectively (Naylor et al. 2001). In the late 80s and early 90s, these increases were driven in part by subsidies for feed, which skyrocketed in the 80s (chap. 8). In the 1990s, diversification to livestock continued to increase despite the reduction in subsidies, probably as a response of farmers to the increasing uncertainties related to the global markets for wheat and other cereals witnessed in the late 90s (Naylor et al. 2001; chap. 8).

Other important changes in cropping systems occurred during this time frame as well, not because of planned policy changes but due to unplanned shocks. Until 1994, the most consistent cropping pattern in the valley was wheat as the winter crop and soybeans as the summer crop. Between 1993 and 1996, whitefly outbreaks decreased soybean area from 69,000 hectares in 1991 to 0 hectares in 1996. One response of farmers was to increase the area planted to sorghum and to summer maize, both of which receive nearly the same level of fertilization as wheat (in contrast to no fertilizer N applied to the N-fixing soybeans), thus doubling the input of fertilizer nitrogen to much of the land surface each year. As noted later, this unplanned change had off-site effects but has not continued in recent years because of the lack of water resources for summer crop irrigation.

Many of these changes—the increasing yields, fertilizer inputs, and animal husbandry—have the potential to influence off-site systems through their effects on nutrient transfers. We evaluated nitrogen loss from current fertilization practices and found that the typical farmer practice results in very large losses of N_2O and NO gas, and large losses of nitrogen from soils in solution, some of which move in the form of nitrate and ammonium from land to freshwater to sea (Matson et al. 1998; chaps. 3 and 10).

Using one of our simulation models (NLOSS, Riley and Matson 2000; Riley et al. 2001; chap. 10), we compared the losses under current fertilizer management to losses that would have resulted from fertilizer application rates used in the valley in the early 80s (180 kg/ha). Assuming equivalent crop characteristics, we estimate that fluxes of N from fields to the atmosphere and water systems increased significantly with the change in fertilizer amount, with NO fluxes more than doubling, N_2O fluxes increasing 10 percent, and NO_3 fluxes increasing nearly twentyfold as fertilizer rates increased from 180 to 250 kg/ha. Thus, intensification through the increased use of fertilizers in the winter wheat crop led to an increase in off-site fluxes of nitrogen. Finally, the transition from the summer soybean crop to maize and other summer crops most likely led to increased N losses, at least for some years—our measurements in summer maize crops indicated that fluxes of N are as large as in the wheat crop. Altogether, intensification of agriculture in the valley clearly has led to increased off-site losses of fertilizer N through a number of different biogeochemical pathways.

Of course, direct changes in land management weren't the only changes experienced by the region through this time period. One of the most critical changes of the past two decades—the amendment to Article 27 of the 1917 Mexican Constitution (the Agrarian Law of 1992)—greatly influenced who could hold and manage the land (Naylor et al. 2001; chap. 8). To remedy the apparent inefficiencies associated with collective land tenure and small plot sizes of *ejido* land ownership (Gates 1993), the reform of Article 27 set up a flexible land-tenure regime inside *ejidos* by allowing communal and individual property holders to rent, sell, and mortgage land legally for the first time. New institutions were established to facilitate this process and to certify official land titles. Joint ventures between the *ejido* sector and the private sector were also encouraged. A legal market for *ejido* land was thus established, replacing the illegal market that was widely acknowledged to exist.

The amendment had important implications for agriculture in Mexico, and especially for farmers in the Yaqui Valley. When the policy was implemented, *ejidos* accounted for just over 50 percent of the total cultivated area, 50 percent of the total irrigated area, and 70 percent of all farmers in Mexico. In the Yaqui Valley, 56 percent of total agricultural area and

72 percent of producers belonged to *ejidos* (CNA 1998; Puente-González 1999). When coupled with changes in government credit and water policies, the reform of Article 27 accelerated increases in private ownership and in the operational size of farms within the valley (Naylor et al. 2001; Lewis 2002; chap. 8), and reduced the number and size of farms that were being managed by *ejidos*. Our analysis suggested that Article 27 and other policy reforms hastened the demise of numerous *ejido* communities as agricultural producers. These changes shifted more land into intensive management, but also had consequences for individuals, in terms of where they live and what they now do. While migration and other population dynamics have not been thoroughly evaluated, discussions with ASERCA (Support and Services for Agricultural Marketing) and other informants suggest that as many as 95 percent of *ejidos* have left farming operations after selling or renting their lands, and many are migrating to the United States or working in areas outside the agriculture district.

Regions Outside the Agricultural District

What are the consequences of these within-irrigation district changes for the ecosystems outside the field system? We'll start with a discussion of changes wrought by biogeochemical or hydrologic transfers, and then move to those more connected to social and economic interactions.

The dramatic increases in fertilizer use and loss to the atmosphere (discussed earlier) should have had significant impacts on air chemistry, pollution, and transport. Our back-of-the-envelope calculations suggested that NO_x emissions were more than enough to drive smog events in Cuidad Obregón, at least during the month immediately following preplanting fertilization (Nov.–Dec.). Air quality in the valley has also been notoriously affected by the burning of crop residues, although burning regulations that levy substantial fines have now reduced the incidence of the practice in the region. Cuidad Obregón has a reputation for respiratory problems, especially in children; agriculture without doubt contributes to that (Kitzes 2005; chap. 8).

In addition, given the large amounts of ammonia and nitrogen oxides emitted during or immediately after fertilization, nitrogen must be being deposited to downwind ecosystems of the Sierra Madre. Nitrogen deposition is known to affect the functioning and species composition in many ecosystems (Matson et al. 1999). There are no deposition monitoring networks in this region of Mexico, and while we several times proposed to carry out the research to track downwind deposition and understand its impact on ecosystems, we were never funded to do so. Thus, the impacts remain supposition.

Substantial amounts of nitrogen also left the individual fields in water flowing through the soil to groundwater systems, off of fields in "tail water" left over from irrigation events, laterally through the soil from fields to the surface water systems, and in effluents from livestock operations and urban centers. There is a tremendous amount about these fluxes that we don't know—again, in part because of unfunded research—but, there is still much that is known. In the surface water systems, high N loading led to some of the highest N_2O fluxes ever measured in freshwater systems, ranging up to 244.6 ng N_2O-N/cm^2/hr (Harrison and Matson 2003; chap. 10). We've estimated that 1,470–2,300 Mg N/yr is transported from the Yaqui irrigation district to the sea each year (chap. 10), where it is linked to enormous phytoplankton blooms (see fig. 10.6). Such euthrophication and blooms in other estuaries have been shown to damage fisheries (NRC 2000; Howarth et al. 2002, Howarth 2008), but the consequences for fisheries or biodiversity of the Gulf of California have not yet been evaluated (chap.7).

In addition to the off-site transfers of chemicals, we evaluated the extent to which expansion of land use into natural lands in the region accompanied the intensification in the irrigation district, and could be related to social or economic drivers. Analyses of satellite time series data and ground-based information indicated a very rapid and relatively recent expansion of land use in the coastal zone, with wetlands, mudflats, mangroves, and other coastal lands being converted to aquaculture ponds at an increasing rate (fig. 4.3; Luers et al. 2006). In addition to the loss of mangroves and wetlands, Pitahaya forests are being cleared to provide land for the aquaculture industry. Pitahaya forests not only provide important native habitat but have unique cultural value for the indigenous people of the region (Luers 2004).

While this expansion is ultimately driven by global markets for shrimp, our analyses also suggested that it was facilitated by reforms in the fisheries, water, and foreign investment laws; changes to the rural credit system; liberalization of domestic and regional agricultural policies, including NAFTA (North American Free Trade Agreement) and PROCAMPO (Direct Farmer Support Program); and constitutional reform of land tenure laws (Article 27), which, as noted earlier, allows *ejido* landholders to rent and sell their land (a significant share of which is coastal land) (Naylor et al. 2001; Luers et al. 2006). On the surface, then, this land expansion may not be associated with activities in the intensive agricultural area, but rather

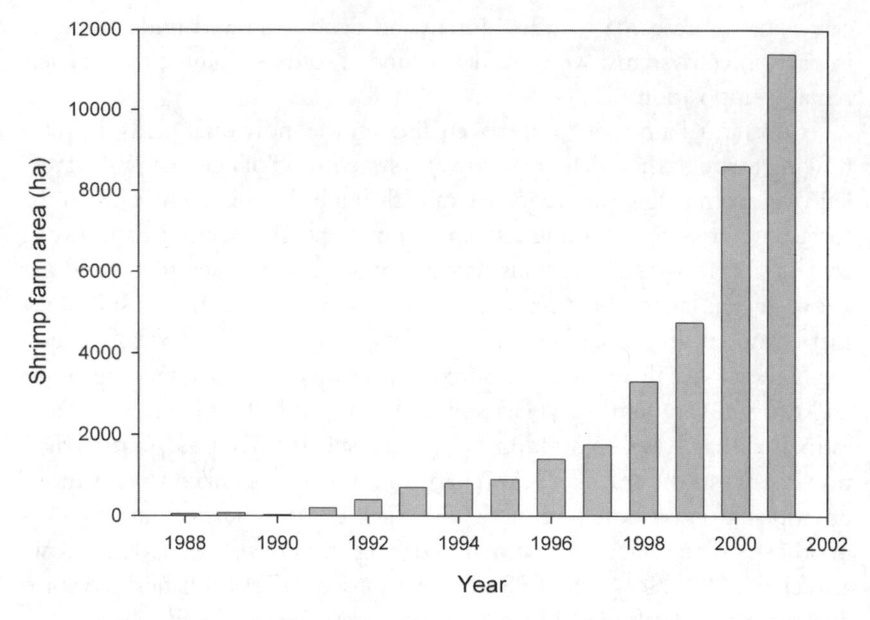

FIGURE 4.3 Shrimp farm area in southern Sonora, 1988–2001.

with forces of globalization and national policies that influence land-use decisions in the region. On the other hand, results from surveys suggest that a significant part of the coastal development is being carried out by landowners from the irrigation district who have the capital or access to credit to allow them diversification of their agricultural activities (Luers 2004).

What about land-use expansion into the Sierra Madres that bound the valley on the east? Analysis of satellite data did not reveal major land conversions taking place in that region—the dry shrubland/woodland systems of the foothills were not being broadly converted to row-crop agriculture, cities, or other uses. Were they, however, being used differently? Were there increasing amounts of cattle management, for example, or other use changes? It was clear that simple approaches of land-cover analysis were not sufficient to tease apart a land change that involves changes in use but not in conversion, and a more sophisticated analysis (e.g., Asner et al. 2003) would be necessary. Unfortunately, as with the analysis of downwind deposition from the valley, we were not successful in garnering funds to carry out the research, so our questions went unanswered. Interestingly, other researchers working in the Sierra Madres uncovered perhaps part of the answer. In surveys of Sierra Madre communities in the region to the east of the Yaqui Valley, Diana Liverman found apparent movement of people (and associated new land uses) from the Yaqui agricultural areas to the foothills (personal communication). Following the change in land tenure laws that allowed *ejido* agricultural land to be rented and sold, and the significant numbers of *ejidatarios* that have taken advantage of the options and are no longer working their lands, it would not be at all surprising to see an exodus from the agricultural district leading to, among other things, increased urbanization and increased land use of the foothills. Further analyses are needed to confirm this as well as the consequences of the moves for the Sierran ecosystems.

As mentioned earlier, an ecosystem loss that occurred at the advent of intensive agriculture in the valley was the loss of freshwater flows in the Yaqui River below the Oviachic Dam, and the loss of riparian habitat there and along the natural streams draining the valley. The development of the irrigation district, however, led to the establishment of many smaller freshwater drainage canals; these were the focus of study regarding loading of nitrogen and other nutrients and transfers from land to the ocean (see chap.10). These canals and altered streams hosted a range of sparse riparian vegetation, including nonnative species such as tamarisk, and supported birds and other wildlife (John Harrison, personal communication). In another lost opportunity, we were not able to carefully evaluate the ecosystem services provided by these altered waterways, their ability to act as refugia or corridors for plant and animal movement throughout the district, or the degree to which they have replaced some of the functions of the Yaqui River, which no longer empties into the gulf.

Finally, a significant land-use change that occurred with intensification of the agricultural district was the rapid urbanization that occurred in the valley over this time. Like many other places around the world, the last forty years have seen an increase in the population, but also the proportion of people living in cities. For example, the population of the municipality of Cajeme has increased by nearly sixfold since 1950. Most of this increase has occurred in Cuidad Obregón, where the population has increased from 31,000 in 1950 to 271,000 in 2005 (INEGI 2005). On the other hand, of the thirty-one localities in Cajeme for which we have a comprehensive dataset spanning 1950 to 2000, twelve have experienced a decline in population since 1950. In the most extreme case, the locality of Daniel Leyva Higuera has seen a decline in population from 299 in 1950 to 6 in 2000 and has disappeared from the 2005 census altogether. Other municipalities such as Guaymas, Etchojoa, and Navajoa also have localities with shrinking populations. Thus, throughout the Yaqui Valley region, larger towns are growing faster than smaller ones, and in many cases, smaller localities are disappearing (fig. 4.4; Matson et al. 2005).

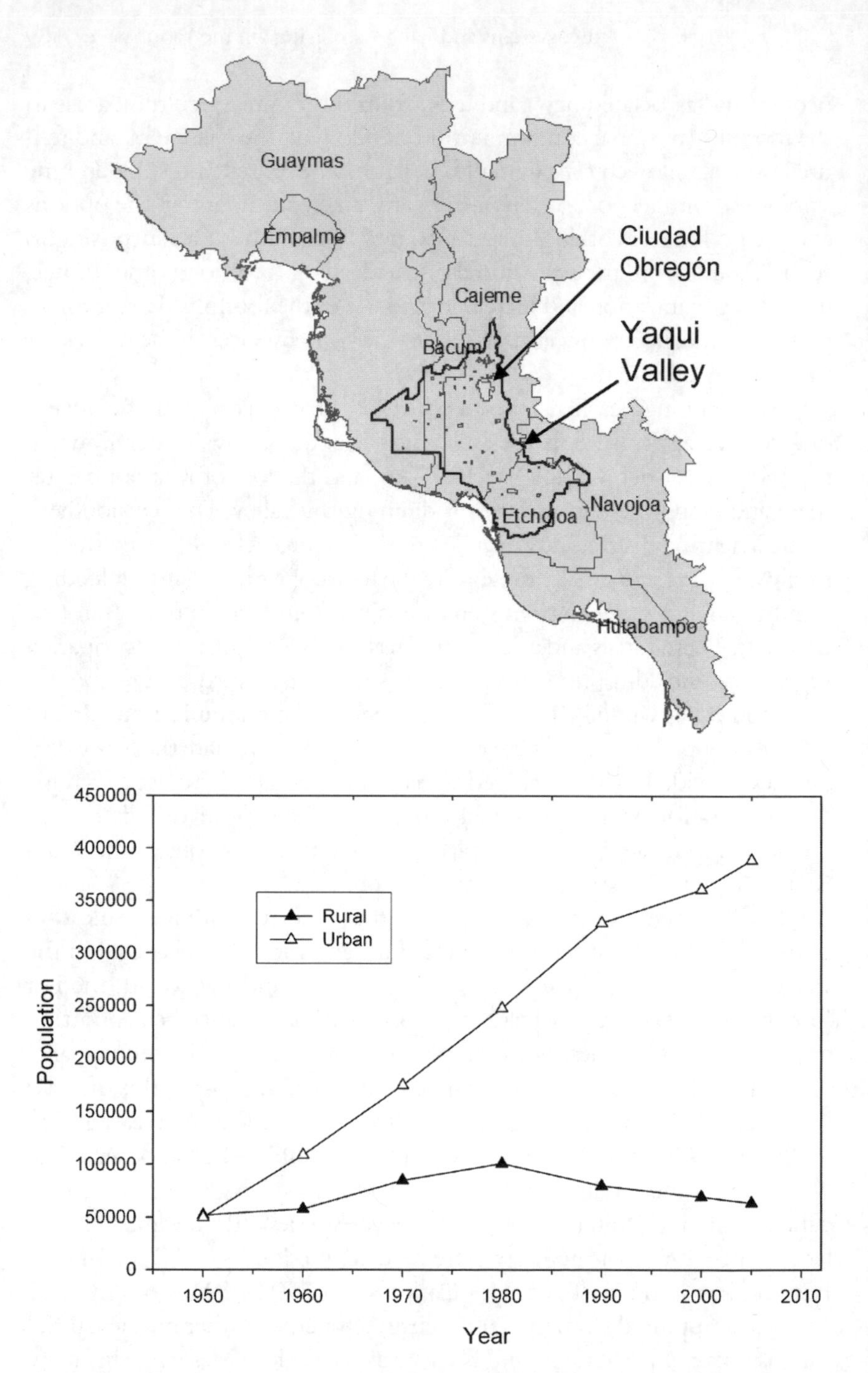

FIGURE 4.4 Population in the Yaqui Valley region, 1950–2005. Data compiled for the municipalities of Cajeme, Bacum, and Etchojoa. *Urban* is defined as any population of 2,500 or more. (INEGI, census years 1950–2005.)

Does Intensification in the Yaqui Valley Spare Land for Nature?

Agricultural intensification in the Yaqui Valley has been remarkably successful in terms of increases in yield and overall food production. That ability to increase yields from around 2 t/ha to 6 t/ha has in theory prevented the need for land conversions elsewhere. At the local to regional scale, indeed, intensification has *not* been accompanied by land conversion for crop expansion, but this fact results from the lack of additional irrigation water. Land changes in the coastal zone, where the greatest action is now occurring, are linked indirectly to the intensive agricultural district, through the investment potential built up by the private farmers of the Yaqui Valley over the past decades. Moreover, the natural ecosystems of the coast and mountains are being affected by intensive agriculture—biogeochemically, hydrologically, ecologically, and economically. While the long-term impacts on those ecosystems are not clear, lessons from other parts of the world suggest that overenrichment of nutrients and removals of freshwater flows, deposition of nitrogen and air pollution impacts, and expanded land use for cattle have significant negative consequences for the native species and ecosystems of the region (MA 2005). It seems likely that intensification is negatively affecting the goal of sparing land for nature in this region.

Where does this leave us? It seems clear to us that intensive agriculture is indeed necessary if we are to protect natural lands and the ecosystem goods and services that they provide to us. But we need a new kind of intensive agriculture—one that efficiently uses inputs to produce high yields, while reducing and eliminating the loss of nutrients and other pollutants, and while protecting the environmental flows of water systems. It is also clear that, to spare land for nature, land-use strategies are needed that purposefully recognize, protect, and conserve critical ecosystems for the ecosystem goods and services they provide us, as well as for the sake of the species with whom we share the planet.

Acknowledgments

The work of a great many researchers has made this chapter possible. We especially thank Lee Addams, Toby Ahrens, Kevin Arrigo, Gregory Asner, David Battisti, Michael Beman, Tracy Benning, Kim Bonine, Esther Cruz, Robert Dunbar, Dagoberto Flores, Steve Gorelick, John Harrison,

Jeff Koseff, Jessa Lewis, David Lobell, Ellen McCullough, José Luis Minjares, Stephen Monismith, Jeanne Panek, Bill Riley, and Peter Vitousek, as well as Yaqui coordinator Ashley Dean, along with Mary Smith and Lori McVay, for facilitating the work of so many researchers in the valley and at Stanford. We thank the NASA Land Cover/Land Use Change Program, the Andrew Mellon Foundation, and the MacArthur Fellows Program for support of the research.

Chapter 5

Linking Knowledge with Action for Sustainable Development: A Case Study of Change and Effectiveness

ELLEN MCCULLOUGH AND PAMELA MATSON

Over the last fifty years, agricultural communities in developing countries have experienced dramatic changes in their requirements for and access to information, knowledge, and know-how related to cropping systems and commodity markets (World Bank 2008; chap. 8). As part of a post–World War II effort to use science and technology to enhance agricultural production, the development and dissemination of *green revolution* technologies (i.e., improved genetic materials and management practices for a range of cereal crops) engaged systems of research, innovation, assessment, development, and deployment to foster goals of agricultural development and economic prosperity (Conway 1997; Hazell 2009). Over time, with the changing landscape of agricultural production (Tillman et al. 2002; Rosegrant and Cline 2003), transformations in global food demand and marketing systems (McCullough et al. 2008), and increasing foci on sustainability objectives including environmental concerns, these systems have been required to evolve, or they ultimately fail to provide relevant support.

The Yaqui Valley provides a perfect opportunity to explore the evolution of knowledge systems for agricultural development. Knowledge systems are networks of linked actors, organizations, and objects that perform a number of knowledge-related functions (including research, innovation, development, demonstration, deployment, and adoption) that link knowledge and know-how with action[1] (Clark et al. in preparation). Included in this definition are the incentives, financial resources, institutions, and human capital that give such systems the capacity to function. While knowledge systems are not the result of master design, they can be at least partially understood and manipulated in ways that improve their performance. In this well-studied home of the green revolution for wheat, we were able to trace some of the changes in the agricultural knowledge system from the early green revolution to current times (McCullough and Matson 2011).[2] Using examples from the Yaqui Valley, we develop hypotheses about the effectiveness of knowledge systems in mobilizing science and technology for sustainable development. We propose that knowledge systems are more agile in the face of challenges and more effective in promoting sustainable development when they critically combine understanding from multiple sources, are managed as results-driven learning organizations rather than pipelines, engage a broad stakeholder community in agenda setting and accountability, and are conducive to evolving as they face different contests and challenges.

Knowledge Systems of the Yaqui Valley

The Yaqui Valley in Sonora, Mexico, provides a unique opportunity to observe the evolution of knowledge systems for agricultural development, to track the most important changes that have influenced farmer decision making over time, and to understand how the knowledge system succeeded and failed under a perpetually changing political, social, and environmental landscape. The Yaqui Valley is home to a key research station of the International Maize and Wheat Improvement Center (CIMMYT),[3] one of the major centers of the Consultative Group on International Agricultural Research (CGIAR) system, and was one of the earliest focal areas for agricultural research and development. Involved in participatory research with agronomists, economists, and breeders, Yaqui farmers and farmer groups helped develop germplasm and management practices that led to dramatic increases in wheat yields. The valley has benefitted directly from formalized systems such as the international agricultural research system, especially CIMMYT, and it has also gained from the national agricultural research system and producer-funded efforts.

Most important, the Yaqui Valley is a place in transition; like many other post–green revolution agricultural regions around the world, it has struggled with a reduction in agricultural extension and decision support, and increasingly must address the off-site impacts of heavy fertilizer use (Matson et al. 1998; Beman et al. 2005; Ahrens et al. 2008), growing competition for water from other sectors (Addams 2004; Schoups et al. 2006b), and maintaining profitability under changing economic and policy conditions (Naylor et al. 2001)—all of which are discussed in other chapters of this book. Such transitions have put new pressures on the Yaqui Valley knowledge system to create and apply different kinds of information at the local level. In response to these pressures, the knowledge system has grown and evolved, though considerable challenges remain.

Using primary information collected through key informant interviews and farmer surveys, as well as literatures and descriptions of the many organizations that are part of the knowledge system, we mapped the structure and dynamics of the knowledge system serving farmers in the Yaqui Valley. Farm behavior and decision making are analyzed from a series of farm surveys conducted by Centro Internacional de Mejoramiento de Maíz y Trigo (CIMMYT) in the 1980s and 1990s (Meisner et al. 1992; Aquino 1998), and subsequent surveys conducted by Stanford University in 1994–95, 1995–96, 2003–04, and 2004–05. In this chapter, we first provide a brief description of the Yaqui Valley, then describe and map connections between the most important players in the Yaqui Valley agricultural system, and finally discuss and evaluate the knowledge to action links that allow actors in the knowledge system to address the critical sustainability challenges that have emerged in recent decades.

A Brief Site Description

As noted elsewhere in this book, the Yaqui Valley comprises 233,000 hectares of intensively cultivated, irrigated lands. The valley is characterized by a semiarid climate and heavy dependence on surface irrigation water during both the dominant winter crop cycle and the summer crop cycle. During the 1960s, semidwarf wheat varieties developed in the Yaqui Valley by CIMMYT scientists and their local collaborators launched the green revolution, which swept across Asia and Latin America in the following decades. Today, roughly five thousand producers farm the valley's wheat-dominated lands, and agriculture remains integral to the region's economy and cultural identity.

Intensive agricultural activity in the Yaqui Valley is associated with heavy inputs of nitrogen fertilizer, freshwater, and agrichemicals (chap. 3; Matson et al. 1998; Ahrens et al. 2008). In turn, agricultural activity impacts surrounding ecosystems via nitrogen runoff that is linked with extensive algal blooms in the Gulf of California (Beman et al. 2005; chaps. 7

and 10); widespread air pollution from the burning of crop residues (Villa-Ibarra and Zepeda 2001); heavy use of agrochemicals including pesticides, herbicides, and fungicides; and the diversion of large amounts of freshwater from alternate uses including natural ecosystems (Addams 2004; chap. 11).

Even as agricultural activity in the valley externally impacts neighboring people and ecosystems, farmers struggle to earn profits in the farm sector. A notable challenge for farmers has been to adjust to a changing policy support structure in the context of many other external shocks (Naylor et al. 2001; chap. 8). At the time of the green revolution, farmers relied heavily on national agricultural institutions for key financial and marketing services as well as production recommendations. Farmers' exposure to fluctuations in input and output markets increased as the state's agricultural support system was scaled back in the 1990s. Land policy reforms comprised a key part of this policy shift and were particularly disrupting to the Yaqui Valley's smallholder peasant farming sector (*ejiditarios*).[4] During the same period as the structural transformation, a series of shocks have rocked the valley—macroeconomic (the peso crisis of 1994), climatic (an eight-year drought culminating in 2004), and biophysical (including invasions of karnal bunt, a disease infecting bread wheat, and the silverleaf whitefly infecting soybeans during the 1990s).

Players in the Knowledge System

The Yaqui Valley knowledge system has influenced how information has been created, adapted, and brought to bear on the challenges that have faced the agricultural sector. Its development and evolution helped to enable the green revolution to take root in the Yaqui Valley and has guided the valley's subsequent response to challenges brought about by the green revolution. Figure 5.1 depicts a circa 2005 snapshot of the Yaqui Valley knowledge system: its organizations, their interactions within the valley and outside of it, and knowledge flows (or lack thereof) between them. Its structure lends insight into which issues gain attention and research effort. The relationships and information flows between the valley's organizations have evolved over the past twenty years in response to the new challenges facing agriculture in the valley, and as the presence of federal institutions has waned. Farmers and the local institutions that serve them have had more decision making pushed down to their level, while their circumstances have grown more uncertain.

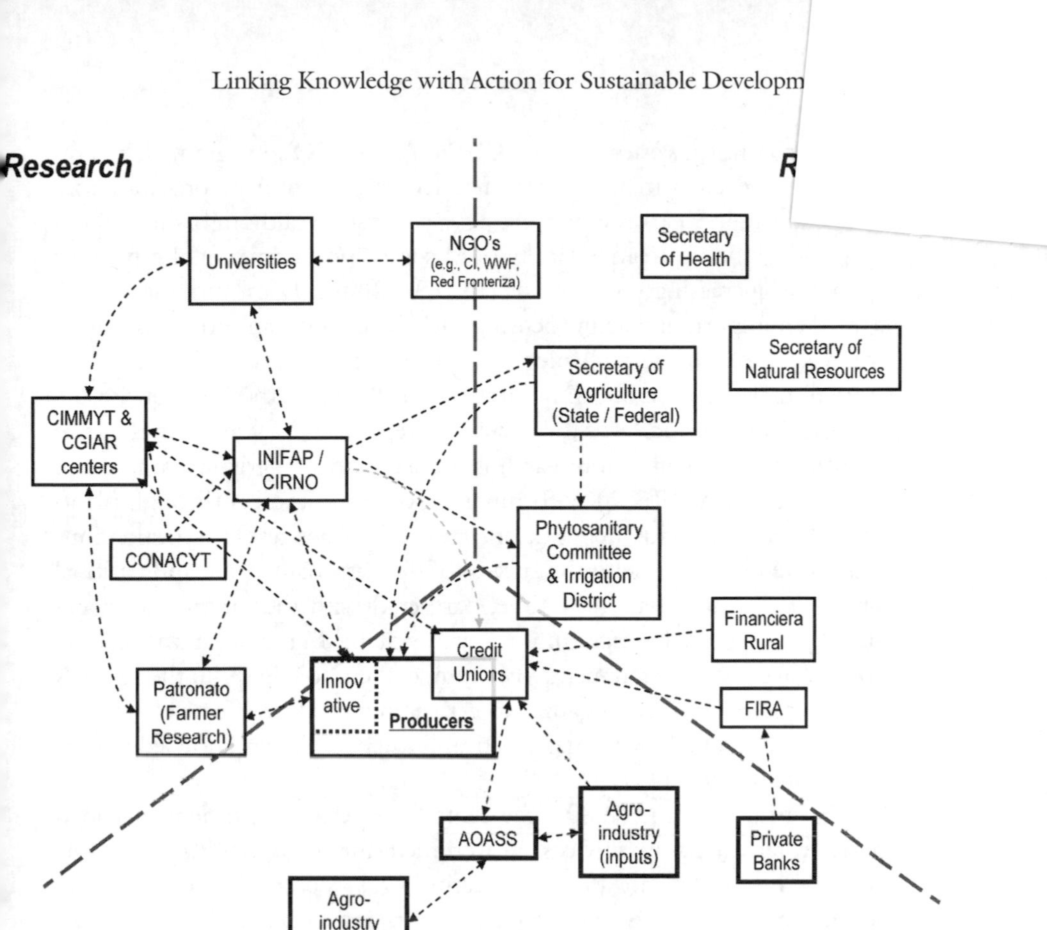

FIGURE 5.1 Knowledge system for agriculture in the Yaqui Valley. The figure represents the key players in the system during 1995–2005 and the linkages between them.

Many of the Yaqui Valley's agricultural research institutions were conceived during the green revolution to support development and delivery of agricultural inputs and services to farmers. These research institutions span the range of international to local. CIMMYT has its home base in Mexico and a chief research station in the Yaqui Valley. Since the 1960s, its most critical role has been to develop improved germplasm, including high-yielding and pest-resistant varieties, for use by farmers in the developing world. Mexico's flagship National Agricultural Research System (NARS) consists of the National Institute of Forestry, Agriculture, and Livestock Research (INIFAP), which operates eight regional agricultural research centers throughout Mexico including one in the valley to serve the needs of Mexico's northwestern states (CIRNO). CIRNO's programs focus on breeding and crop management. The federal government provides core budgetary support to the national research system, although supplemental project funding from a variety of Mexican and international funders is becoming increasingly important (INIFAP 2004). Local funding has assumed an important role in focusing the efforts of organizations conducting research in the Yaqui Valley on problems prioritized by farmers. The Agricultural Research and Experimentation Board of the State of Sonora (PIEAES, known locally as the *Patronato*) represents all farmers in the state of Sonora on agricultural research issues, supporting agricultural research and development (R&D) with funds raised from farmers through planting fees. A nationwide network of state-level farmer research foundations was created in response to the success of the Patronato. State and national university research has, over time, partnered with the international, national, and regional agricultural research system. Many organizations, on paper, have mission statements that compel them to engage in the region's knowledge system; the organizations that engage most effectively rely on dynamic individuals who make critical linkages between knowledge producers and users.

In the 1970s and 1980s, the central Mexican government provided agricultural extension services. Government-funded agricultural support declined during the late 1980s and early 1990s as part of sector-wide policy reforms. According to both practitioners and scientists, the downscaling and decentralization of federally financed extension brought a marked decline in official linkages between science and technology producers and their users. Delivery of extension services was subcontracted to private sector providers, and fee recovery was instituted for farmers who could afford extension (Umali and Schwartz 1994). Some of these functions were replaced, more or less, by technical advisors employed by the Yaqui Valley's credit unions. Their extension focus is dictated by farmer demand and credit union interest; a mechanism no longer exists for provision of extension services motivated by anything other than raising a farmer's private profitability.

Most of the Yaqui Valley's farmers are organized in credit unions—farmer associations that provide members with credit, seeds, fertilizer, other inputs, insurance, postharvest storage, marketing, and technical assistance (fig. 5.2). Farmers must operate a relatively large amount of land to qualify for membership in these powerful associations (100 hectares minimum[5]). While a few credit unions and similarly conceived public organizations cater specifically to smallholders and farmers from the *ejida*

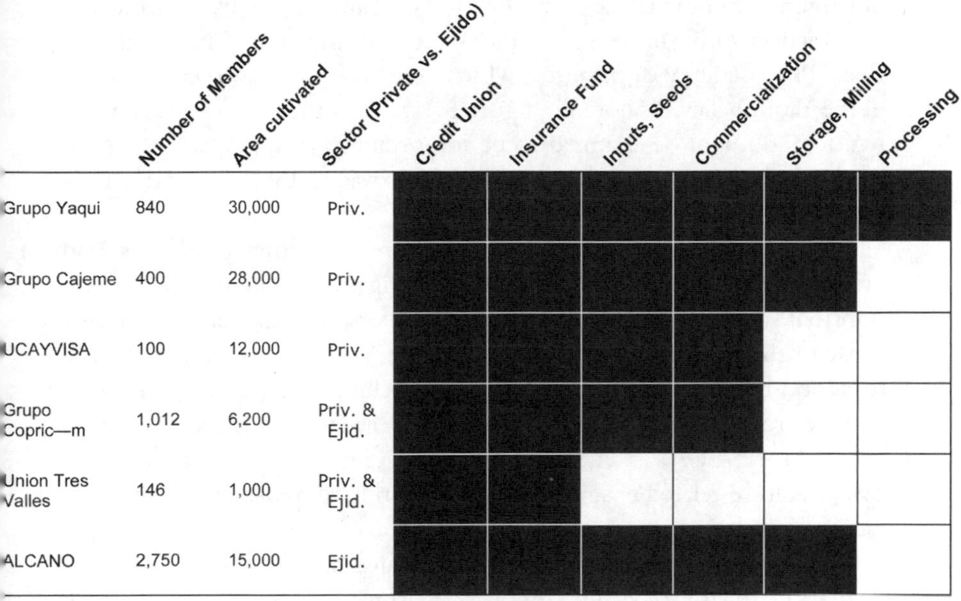

	Number of Members	Area cultivated	Sector (Private vs. Ejido)	Credit Union	Insurance Fund	Inputs, Seeds	Commercialization	Storage, Milling	Processing
Grupo Yaqui	840	30,000	Priv.	■	■	■	■	■	■
Grupo Cajeme	400	28,000	Priv.	■	■	■	■	■	
UCAYVISA	100	12,000	Priv.	■	■	■	■		
Grupo Copric—m	1,012	6,200	Priv. & Ejid.	■	■	■	■		
Union Tres Valles	146	1,000	Priv. & Ejid.	■	■				
ALCANO	2,750	15,000	Ejid.	■	■	■	■	■	

FIGURE 5.2 Major credit unions: characteristics and services provided.

sector, they are typically less powerful and offer fewer services. Smallholder peasant farmers, in general, are more isolated from the agricultural R&D system in the Yaqui Valley. Credit unions play a key role in scaling up adoption of new technologies that have gone through development and testing by research organizations and farmers (see following paragraphs on innovative farmers). They draw on agronomy, market information, and agribusiness outreach materials to inform management recommendations for their members. Credit unions also provide insurance payments in the event of crop failure to members who follow official management recommendations. Many farmers adhere to the official recommendations because they are considered a requirement to receive credit, insurance, access to markets, and other benefits of the credit unions. A federation of southern Sonora's credit unions, known as AOASS, is actively engaged in local, state, and even national policy processes, typically centered on securing minimum support prices for key commodities.

In addition to the credit unions, two local organizations operating under government charters with funding through planting fees, are important determinants of farmers' irrigation and pest management, respectively. These are the Yaqui River Irrigation District, which oversees maintenance of irrigation infrastructure and allocation of water to groups of farmers (in conjunction with Mexico's National Water Commission [CNA]), and the Local Phytosanitary Committee, which guides pest management strategies in the Yaqui Valley. As noted in figure 5.1, many other agencies exist in the region and might be assumed to be important in this knowledge system, but because they are not connected to key decisions, they are not described here.

At the center of the knowledge system are agricultural producers, both private landowners and *ejiditarios*. According to our surveys, farmers report that they draw heavily upon their own personal experiences, as well as those of their family, friends, and neighbors, for background information to support field management decisions, including fertilizer application.[6] Farmers report that informal sources of information, such as the experiences of their friends and neighbors, are most important to them, followed by agriculture-related programming found in local newspapers and on the radio.

Given the potential importance of this informal network for sharing information, the role of leading innovative farmers is especially important; in our analysis and experience in the valley, a number of individual producers emerged as critical players. These self-described innovative producers typically have college or graduate educations and are recognized by researchers and organizations in the region as influential key decision makers.[7] They often collaborate with researchers from national and international institutions to develop and test new agricultural management technologies. They also often hold influential roles in or have influential relationships with credit unions, farmer-sponsored and government-sponsored research organizations, and political offices. Innovative producers form a critical link between researchers and the farming community, and as such play a vital role in validating new technologies, assessing their economic profitability, and influencing the research agenda by articulating farmers' needs and priorities to researchers.

In today's Yaqui Valley, science and technology are harnessed for improved field and farm management, not through a "pipeline" that provides information from researchers to farmers in a one-directional flow, but instead through a collaborative process by which farmers guide researchers' focus and validate new technologies. Farmer testing and validation is critical to eventual adoption of new technologies. Producers typically are motivated to enter into trials for technologies aiming to raise or stabilize their yields or lower their production costs; trials of management strategies to decrease off-farm impacts are not prioritized, although such outcomes may result from yield or profit improvements. Innovative producers assume the risks of crop failure on plots planted to research trials, though they are affiliated with credit unions. These producers provide credit unions access to key data points around new technologies that allow credit unions to understand them and pave the way for their eventual incorporation into mainstream management recommendations.

Evolution in the Functioning of the Knowledge System

The knowledge system's effectiveness at harnessing science and technology for improved field management, and its ability to evolve and respond to changing conditions, can be illustrated through characteristic examples. Here, we evaluate the knowledge system's contributions to improved decision making in the face of typical green revolution and post–green revolution challenges: transfer of new crop varieties, planting system innovations, fertilizer and water management, and crop diversification.

Transfer of New Germplasm

The green revolution is most often associated with deployment of improved germplasm along with the packet of inputs and management information accompanying new varieties. The Yaqui Valley breeding system has been effective in reaching farmers with new wheat varieties. CIMMYT focuses on upstream breeding of lines that perform well over a diverse range of agroclimatic conditions. Mexico's NARS continues to develop lines that are promising for Mexican farmers, and the Patronato receives breeder quality lines from the local NARS station and sells them to credit unions who multiply them and eventually sell certified seeds to farmers. This system for wheat germplasm improvement is consistent with CIMMYT's global model, though Yaqui Valley farmers enjoy particular proximity to one of CIMMYT's most important research stations and the regional NARS headquarters.

This system has evolved and been tested over time, and works remarkably well for Yaqui Valley farmers, who often adopt new varieties almost as soon as they are released by CIMMYT (see Naylor et al. 2001). As illustrated by a recent durum wheat leaf rust outbreak, Yaqui Valley farmers are well positioned to deploy disease resistant wheat varieties in short periods of time. In early 2001, scientists detected durum wheat leaf rust in the Yaqui Valley. CIMMYT and NARS researchers immediately accessed resistant races from their seed bank, developed, multiplied, and released a resistant, high-yielding, advanced line which was broadly adopted by valley farmers less than two years after the outbreak (Camacho-Casas 2003). The Patronato plays a critical role in linking farmers to international, national, and regional germplasm improvement efforts. No analogous structure exists for linking farmers with improved management recommendations produced by CIMMYT, Mexico's NARS, or other researchers.

Planting Systems

The raised bed planting system spread to farmers through innovative farmers who were linked into the local agricultural R&D system. During the 1980s and 1990s, most wheat farmers in the valley shifted from flat bed to raised bed planting systems. Raised beds have been used in high moisture systems to improve soil drainage for millennia, but the Yaqui Valley was the first semiarid, irrigated region to scale up adoption of the planting system (Sayre 2003). Following successful NARS trials that were replicated at a local university in the early 1980s, innovative producers in the valley began to test the raised bed system in collaboration with national and international researchers. This system allowed for more efficient irrigation at lower effort than the flat field alternative, while facilitating lower planting density, more effective weeding, higher yields, and more efficient fertilizer uptake (Sayre and Moreno Ramos 1997). The new planting system required only minor adjustments to existing machinery. In surveys, farmers attributed their adoption to improved water management, better weed control, and easier field access with machinery (Aquino 1998).

Early innovative adopters received good results while also lowering their production costs, and soon their relatives, neighbors, and friends began to adopt the technology. By 1991, over 50 percent of farmers in the valley had adopted the system (Meisner 1992). These adopters, mostly members of credit unions, were spurred by credit union endorsement of the raised bed system. Small farmers from the *ejidal* sector were the last to leverage the technology, though by 2004, a farmer survey confirmed that over 90 percent of valley farmers had adopted it. Agronomists from all over the world have visited the valley to participate in CIMMYT's intensive raised bed training program, carrying their understanding back to farmers as far as Central, South, and East Asia. This example underscores the importance of innovative producers in spanning the boundary between agronomic research and adoption of improved farming practices. Researchers focusing on the raised bed planting system were well connected to innovative producers, and they were able to identify a technology that resonated with them and enhanced their on-farm profitability.

Fertilizer Management

Starting in the 1960s, the Mexican federal government began subsidizing fertilizer production, keeping domestic fertilizer prices below international prices (Naylor et al. 2001; chap. 8). These economic incentives, along with extension recommendations that encouraged heavy rates of fertilizer application, resulted in a continuous increase in fertilizer use intensity from the 1960s, to an average rate of 250 kg/ha (Naylor et al. 2001). The majority of fertilizer is commonly applied one month before planting, a practice that, in combination with the heavy rates of application, has been linked with large losses of fertilizer from fields to the atmosphere and watersheds, and with associated economic losses for farmers (Matson et al. 1998; chap. 3). In the late 1990s, our multidisciplinary research team investigating off-site impacts of intensive fertilizer use recorded some of the highest ever reported measurements of gas fluxes from farmers' fields. Later, high levels of nitrogen in streams and canals draining the valley were associated with high nitrogen trace gas emissions, and phytoplankton blooms in the Gulf of California were linked with fertilizer runoff associated with major irrigation events in the Yaqui Valley (chap. 10).

Starting in the late 1990s, a number of factors led to greater interest in increasing the efficiency of fertilizer use. These include a growing awareness of the magnitude of fertilizer losses and their associated environmental impacts; scaling back of fertilizer subsidization and public extension; currency exchange rates, which raised the effective price of fertilizer for farmers; and a sustained decline in global wheat prices (Naylor et al. 2001). The economic motivation for increasing the efficiency of fertilizer use was potentially an important one for the agricultural community, as surveys in the late 1990s indicated that fertilization was the single most important cost component in Yaqui Valley farm budgets. Stanford University and CIMMYT researchers, with funding from US federal funding agencies and foundations interested in both agricultural development and sustainability objectives, began to evaluate several alternative fertilizer management strategies that decreased the total N applied and timed applications more closely with plant uptake. All of the alternatives resulted in some reduction in N losses, and the "best" alternative, which called for 25 percent less N per hectare, to be applied during and after planting, resulted in dramatically lower losses of N. Without differing in yields or grain quality, the best practice resulted in estimated cost savings to farmers equivalent to 12–17 percent of after-tax profits (Matson et al. 1998).

Next, the research team tested the apparent win-win practices in farmer fields, collaborating with innovative producers. Despite successful on-farm trials and farmer workshops to share the new approach, surveys indicated that application rates did not decrease (see chaps. 3 and 9). Inquiries into fertilizer application decisions revealed that lack of adoption of the new management approaches was likely due to uncertainty about soil and climate conditions by both farmers and credit unions (Lobell et al. 2004), especially related to the risks of insufficient rates and late timing of fertilizer applications for optimal crop response. In some years, farmers could face greater risk of low yields with the best practice, especially if late rains render fields inaccessible to machinery, delaying the second fertilizer application beyond the point of optimal plant response (Avalos-Sartario 1997). According to surveys, farmers managing smaller plots were less deterred by this risk and more likely to switch to no preplanting fertilizer than farmers managing larger plots (Aquino 1998).

After farmers reported that climate uncertainty was a key factor in determining the timing and rates of fertilizer application, researchers began to explore the topic in more depth. Stochastic modeling of yield variability using historical climate data indicated that uncertainty about soil and climate conditions could at least partially explain the observed fertilizer overapplication—farmers' profits are more sensitive to peaks incurred by yield increases due to favorable climate conditions in good years than they are to troughs incurred by excessive fertilizer application the rest of the time (Lobell et al. 2004). Farmer interviews supported these findings. Even though common practices were more expensive than best practices in average years, preplanting overapplication helps farmers protect against crop failure in very wet years and prevents them from forgoing the opportunity to profit from highly favorable climate conditions. Meanwhile, credit unions in the valley were advising that farmers continue the high-input, low-risk, time-tested management practice rather than adopt the newly developed best practice. The technical units of the credit unions reported that advice was based on concerns about loan recuperation rates under potential crop failure across the range of farms, soils, and managers that they serviced. Because credit unions also serve as retailers of fertilizer, seeds, and other inputs, they had some financial incentive to maintain input sales volumes.

An approach to addressing farmers' production tradeoffs in adopting improved fertilizer practices has been to improve soil and climate information in order to allow farmers to better synchronize fertilization rates and timing with plant demand. This could simultaneously increase profitability while decreasing nitrogen pollution in adjacent air and water systems.[8] Under the leadership of CIMMYT agronomist and team member Ivan Ortiz-Monasterio, an in-field nitrogen diagnostic tool was developed to provide farmers with real-time information on soil N availability. This sensor-based tool allows farmers to monitor nitrogen and add more if needed, so they avoid overfertilizing in most years without forgoing their full yield potential in optimal climate years. After calibrating the sensor in experimental plots, Ortiz-Monasterio began trials with innovative producers and then approached the Patronato, which engaged credit unions in the trials. Credit unions agreed to fund half the costs of eight sensors, and Fundación Produce Sonora funded a proposal for the other half. By engaging the credit union technical units, the researchers brought the toughest critics on board from the beginning, allowing them to reach a much broader audience of mainstream farmers (chap. 9). The sensor technology is now increasingly used in the Yaqui Valley and is spreading into other areas of Mexico and Asia.

Crop Diversification

Since the valley's inception as a green revolution site in the 1950s and 1960s, wheat has dominated the cropping system, thanks to favorable biophysical conditions for its cultivation, locally developed breeding innovations, and a generous, reliable endowment of surface irrigation water. Cotton, soybeans, and maize have all, successively, played important roles during the spring–summer crop cycle, which has been left largely vacant since 2002 due to drought. In recent years, the strategy of wheat-dominated agriculture has been called into question due to the decades-long trend of declining wheat prices, rising production costs, and concerns that wheat-dominated agriculture is virtually impossible without widely available cheap water. State and federal policy makers, along with progressive farmers, have begun to explore and promote crop diversification as a strategy for stimulating renewed growth in the Yaqui Valley's economy while using water more productively.

Knowledge and information (related to production and marketing) are central to diversification, though it has been unclear that the Yaqui

Valley knowledge system, focused for so long on a single commodity, could evolve to support the knowledge needs. While many farmers in the Yaqui Valley have cultivated a diverse set of grains and oilseeds, only 33 percent have ever expanded their cropping portfolios beyond basic grains into fruits, vegetables, and potatoes.[9] Our surveys and interviews show that, for Yaqui Valley farmers, agricultural diversification is indeed knowledge-limited, but not on the production side. Farmers claim that, rather than lack of production know-how, uncertainties in market opportunities and high production costs prevent them from diversifying. Information barriers to developing marketing opportunities and negotiating favorable terms of sale can be formidable, especially for high-value, highly perishable products whose markets are characterized by their high transaction costs, exacting quality and safety standards, and resulting production and marketing risks (see Joshi et al. 2007).

Yaqui Valley knowledge institutions are not currently poised to provide the market information and logistics support that can enable crop diversification. Interventions are needed to lower transaction costs via development of business capacity, support for forward contracting, and facilitation of marketing opportunities. For example, producer associations could negotiate contracts with buyers, coordinate supply logistics, and provide credit, technical assistance, and market information. The question remains whether the green revolution–initiated, wheat-focused knowledge system can evolve to meet these needs.

Water Management

The Yaqui Valley's heavy reliance on irrigation water is perhaps its most defining characteristic. The recent drought was a sobering reminder of the importance of effective water management decision making and productive water use. During the 1990s, management of water resources was decentralized from the National Water Commission (Comisión Nacional del Agua or CNA) to water user associations (Johnson 1997). The Yaqui River Irrigation District was one of the first to be transferred to users nationwide, due to its reputation for promoting productive agriculture. In 1992, the district took responsibility for operation and maintenance of irrigation canals and infrastructure, and a number of deep wells in the valley (Addams 2004). The National Water Commission continues to operate the Yaqui River reservoirs and negotiates with the irrigation district to allocate reservoir water to crop irrigation at the start of each year's wheat growing season (chap. 11).

Following decentralization of water management, a long string of dry years led the irrigation district to make a series of high-risk reservoir allocation decisions, based on increasingly optimistic expectations of future in-season runoff (Addams 2005). In 2002–03, after about six dry years, and with reservoirs at precariously low levels, the irrigation district authorized the sale of planting permits for an area that was 40 percent larger than it should have. According to long-term climate data, there was only a 50 percent chance that there would be enough reservoir inflows in season to allow the district to supply water to the fields whose planting it had authorized (Addams 2005). The inflows did not occur, and the irrigation district scrambled to procure equipment to pump water out of the main reservoir's dead storage capacity in order to deliver on the planting permits it had sold farmers. The district entered the following growing season with empty reservoirs. That year, the aggregate value of the Yaqui Valley's agricultural output plummeted to M$383 million, less than 40 percent of the average output value during the preceding decade in real terms (McCullough 2005).

After the Yaqui Valley's drought crisis illustrated to local water managers the consequences of their optimistic allocation decisions, the irrigation district has begun to adopt a more conservative approach to water management (Jacobs et al. 2010). This extreme event illustrated to local water managers that their allocation decisions had been far too optimistic, which led to evolution of the knowledge system. The irrigation district worked with scientists at the CNA, Stanford University, and elsewhere to develop a water management strategy with risk-based operating rules that accounted for inflow uncertainty (chap. 11). A modeling and stakeholder deliberation effort culminated in adoption by Mexico's CNA and the Yaqui Valley's irrigation managers of surface water allocation rules designed to maintain minimum reservoir outflow levels through extended periods of variable and unknown inflows (Minjares 2004). The modeling effort is focused on an improved understanding of the entire hydrological-geological-agricultural system, in order to develop strategies for improved conjunctive management of surface and groundwater (Schoups et al. 2006b).

Since farms comprise the largest source of demand for water in the Yaqui Valley, improved water management at the farm level will be important. As with fertilizer, powerful incentives provided by the credit unions and the irrigation district's planting permit system enforce farmers' water use levels (McCullough 2005). Raising the price of water alone is unlikely to change practices. Even after the drought in 2003–04, when the volumetric price of water doubled, irrigation records confirm that farmers who received water applied the same volumes as they did when it cost half as much (McCullough 2005). With improved data on crop and yield responses to different irrigation management practices, and buy-in from credit unions, it might be possible to reduce the irrigation district's allocation volumes, or to provide other incentives for more productive farm-level irrigation practices.

What Can Be Learned from the Yaqui Valley Knowledge System?

The Yaqui Valley is subject to its own special combination of organizations and individuals who face very place-specific challenges that change over time. While it is impossible to generalize to the world based on analyses of just one place, our range of experiences in linking knowledge to action in the valley provide some take-home messages that are worth sharing. Moreover, as part of the broader "Knowledge System for Sustainable Development" study, they have been generalized more broadly (Clark et al. in preparation). In the following paragraphs, we summarize some of the learning that has emerged from the examples described in this chapter.

Knowledge systems are more likely to be agile and effective in promoting sustainable development to the extent that they critically combine understanding from multiple sources.

The Yaqui examples illustrate the critical role of many different organizations and individuals in the knowledge system. This knowledge system is a multiconnected system with multidirectional information flows, and it integrates new scientific knowledge along with experiential knowledge and the knowledge built through decades of the agricultural tradition in the Yaqui Valley. It engages information and input from researchers of many different kinds (e.g., agronomists, economists, biogeochemists, climate scientists, hydrologists, geographers, ecologists, etc.). This knowledge arises from activities in international, national, state, and local research organizations, some of which is interdisciplinary in form. It also engages the knowledge and know-how of managers and farm advisors from state and private farm organizations, including credit unions. Most important, it relies on the knowledge of producers—innovative, risk-taking farmers as well as those who are ruled by their credit unions—who draw on their experience as well as the knowledge they purposefully coproduce with others in the research endeavor. For example, researchers were able to understand how farmers' nitrogen management practices were optimized over a multiyear rather than single-season time horizon through farmer interviews, close collaboration, and analysis of plot-level data. This understanding was essential for designing and testing better nitrogen management practices that farmers were likely to adopt. None of the examples discussed in this chapter would have succeeded without the knowledge, input, and feedback from many key players, and indeed, the agility of the system to respond to new challenges is based in part on the range of players.

Not surprisingly, communication and translation across perspectives and players are key to the workings of this system and many others (Cash et al. 2003). We encountered communication challenges within the research community itself, as we strove to contribute knowledge that was salient to the Yaqui Valley community yet met our own, often very different, research motivations and perspectives. We also struggled with these challenges as we sought to bring together farmer and researcher and farm advisor perspectives. Respect, ultimately, was the essential characteristic that enabled some of the integration.

Knowledge systems are more likely to be effective in promoting sustainable development to the extent that they are strategically managed, not as pipelines for transferring R&D to users, but rather as results-driven learning organizations.

As noted, the Yaqui knowledge system engages many different organizations and individuals; in the linkages and in the functioning of the information system, the Yaqui case negates the simplistic view of the one-way flow of scientific information from the agricultural research community to the user community. The pipeline model does not work in the Yaqui Valley. Instead, considerable two-way discussions motivate the scientific endeavor, with the research community not only providing improved approaches (sometimes motivated by concerns other than those of the user community), but also being sensitive and responsive to the needs and challenges that are faced by the user community and that impede their success. Iterative feedback and learning between the user and research communities is key to the function of the system.

Other research has identified the importance of effectively managing the boundaries between scientific knowledge and decision making (Guston 1999; Cash et al. 2003; NRC 2006). Whether through formal organizations (or units within organizations) that are designed to act as intermediaries between the science and decision-making communities, or through individuals who explicitly assume that role, the boundary-spanning function works best if there are clear lines of responsibility and accountability to both sides of the boundary, and if they provide venues for interaction or coproduction of knowledge by actors from both sides, often through the development and use of *boundary objects*, such as models, scenarios, maps, or other outputs (Cash et al. 2003). The Yaqui Valley examples illustrate the importance of boundary organizations and individuals and how they evolve.

State-funded agricultural extension systems are a prime example of boundary organizations, with extension agents connected strongly to both the academic research world and to the world of agricultural decision makers (Clark et al. in preparation; Cash et al. 2003). As extension disappeared in the Yaqui Valley, credit unions began to replace that boundary role. As trusted farmer associations that also provided credit lending, the credit union farm advisors became an important conduit for new agronomic information as well as information about commodity and input markets. Since innovative farmers are viewed as influential leaders in these unions, they provided a clear conduit for feedback from the farming community. However, credit unions ultimately lacked the responsibility to link to the research community (other than those they were funding), which constrained the flow of information and feedback from farmers to the research community. The credit unions' financial stake in the management strategies of farmers, and their inherent risk aversion, also prevented them from motivating, on their own, innovations that would bring about economic and environmental benefits for farmers and the region.

A number of individuals also played extremely important boundary-spanning roles in the Yaqui Valley. Foremost of these was Ivan Ortiz-Monasterio, team member and agronomist at the CIMMYT field station located in the Yaqui Valley. His boundary role was not mandated by CIMMYT—far from it. Rather, he developed a unique role—one that would not be automatically replaced should he leave the valley—that linked the research community (including the Stanford team, local university researchers, and government researchers) with the innovative producers. Because of mutual trust and credibility that had been built up through experience with farmers in the valley (via professional and social interactions), Dr. Ortiz-Monasterio was a conduit for ideas, approaches, and feedback to and from the farmers and the research community; similarly, because of trust and credibility in the world of agricultural sciences and sciences more broadly, and because of the especially high respect and value placed on him by the research team, he was highly influential in determining the foci of the research team, including its outreach and dissemination activities. As the fertilizer example makes clear, the development and testing of new management approaches that make sense environmentally and economically in the Yaqui Valley is a result of multilinked and iterative interactions among the research community, innovative farmers, credit unions, and farmers in general, often with Dr. Ortiz-Monasterio at the center.

Knowledge systems are more likely to be effective in promoting sustainable development to the extent that they engage scientists, users, and society in deliberation of agenda-setting and accountability.

The agenda of the Yaqui Valley agricultural region has been for many decades a green revolution agenda, focused on improvements of yields of primarily cereal crops. Indeed, until the policy changes of the 1980s and 1990s, the agenda was much more about yield than about profit, and the idea that biggest yields are best (and greatest inputs are as well) remains a prevalent perspective. Not surprisingly, the producers, credit unions, and agricultural research organizations were and still are the agenda setters within the valley, and external funding sources throughout the green revolution years extending through the end of the century were primarily concerned with agricultural development. The Yaqui Valley's strongest institutions serve the "typical" farmers' interests quite well, often at the expense of others. The poorest social sector farmers and agricultural laborers, who function outside of credit unions or who are members of weak ones, are not well connected; neither are nonagricultural interest groups. Some of the *ejido* peasant farmers have little voice, and their concerns are not likely to emerge in the agricultural research or social agenda of the valley.

With increasing challenges in the agricultural sector as well as increasing competition for water, for environmental protection, and for the protection of human health, voices of other groups in the region are beginning to be heard. Likewise, the influence of nongovernmental bodies and groups from outside the region, including the funders of and participants in research, is changing the agenda. Research that develops and presents new kinds of knowledge (e.g., on air and water pollution or childhood health issues), and the development of new boundary objects directly related to fertilization and irrigation in the valley, such as the image of phytoplankton blooms in the Gulf of California (see chap. 10, specifically fig. 10.6, for details) can change the dialogue and the agenda of the knowledge system. One small indicator of this slow change can be seen in the title of farmer workshops held by our research team: in the early years of the study, workshops were titled "Increasing the efficiency of nitrogen fertilizer use," while one recent one was titled "A transition to sustainability in the Yaqui Valley." Indeed, one of the main contributions of this project may have been to move the agenda toward one that includes environment and sustainability issues, thus setting the stage for further change.

Knowledge systems are more likely to be effective in promoting sustainable development to the extent that they evolve and change with the changing demands and altered political, social, and environmental environments in which they operate.

All of the preceding conclusions add up to the point that knowledge systems are not static; indeed, if they are, they will fail. Moreover, our analysis suggests that if the research community is interested in connecting the production of knowledge with decision makers and decision making, it needs to recognize the dynamics of the knowledge system and to link purposefully to it in strategic ways that may change over time. The Yaqui Valley experienced a rapid and dramatic devolution of decision-making responsibility to the organizations and individuals within it and influencing it during the 1990s and 2000s. This period of devolution coincided with a number of key shocks, which forced the knowledge system to grow and evolve, and it did so with considerable success. It remains to be seen if it is flexible and agile enough to successfully make the transition to completely new cropping systems and resource management approaches that may be needed to meet the sustainability challenges of the coming decades.

Acknowledgments

This work was supported by a grant from the National Oceanic and Atmospheric Administration's Office of Global Programs for the Knowledge Systems for Sustainable Development Project (see http://www.ksg.harvard.edu/kssd) and by a grant from the Packard Foundation to the Center for Environmental Science and Policy in the Stanford Institute for International Studies. The authors would like to thank Wally Falcon, Roz Naylor, Ivan Ortiz-Monasterio, David Lobell, Lee Addams, Dagoberto Flores, José Luis Minjares, Louis Lebel, Peter Jewett, and Monika Zurek. In particular, we would like to thank William Clark for inspiring this work.

Chapter 6

Exploring Vulnerability in the Yaqui Valley Human-Environment System

PAMELA MATSON, AMY LUERS, AND ELLEN MCCULLOUGH

The vulnerability of people and places has emerged in the past decade as a central concern related to climate change (IPCC 2001; IHDP 2001) and a key question in the emerging field of sustainability science (see Clark et al. 2000; Kates et al. 2001; Turner, Kasperson, et al. 2003; Turner, Matson, et al. 2003; Eakin and Luers 2006). As it becomes increasingly clear that not everyone or every place is equally vulnerable to the rising frequency of environmental stresses, such as floods and drought resulting from climate change, or to major price shocks and policy changes, researchers and practitioners are seeking to understand what makes some places more vulnerable than others. What can be done to reduce the likelihood of harm for the most vulnerable? How can more resilient and adaptive communities and ecosystems be built or managed?

To address these questions, many are turning to integrated assessment approaches that address the vulnerability of coupled human-environment systems. These consider both the natural characteristics and resources (sometimes called *natural capital*) and the human and social characteristics (called *social capital*) of a system in light of the impacts, responses, and outcomes of one or more stresses acting on it (fig. 6.1; Turner, Kasperson, et al. 2003; Eakin and Luers 2006). Such perspectives on vulnerability require that we ask not just what the stresses and their impacts are, but how the specific characteristics of people and places influence their responses and their likelihood of suffering harm. How are such changes and their

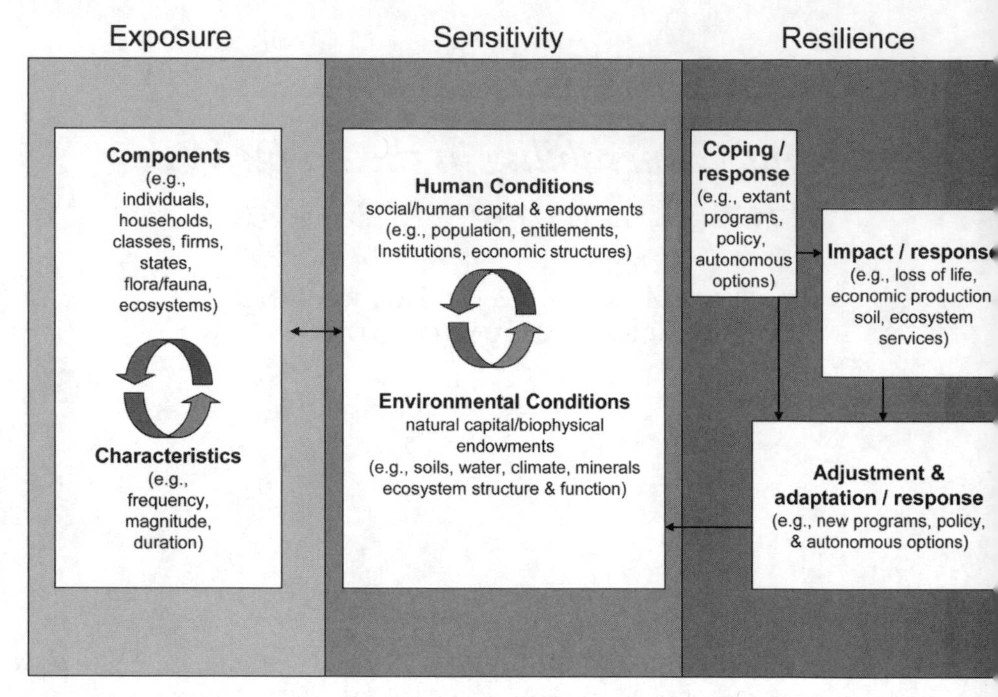

FIGURE 6.1 Integrated vulnerability approach (Turner et al. 2003a). © 2011 by the National Academy of Sciences, USA.

consequences attenuated or amplified by different human and environment conditions? What happens when these stressors interact?

The Yaqui Valley has provided a useful though incomplete (and often unfortunate) set of case studies with respect to the study of vulnerability. Our experience with, and our understanding of, vulnerabilities in the region were critical in the development of a framework for the analysis of vulnerability published in the *Proceedings of the National Academy of Sciences* in 2003 (fig. 6.2; Turner, Kasperson, et al. 2003). This framework suggests that vulnerability is best understood through the lens of *coupled human-environment systems*, where vulnerability and sustainability are a result of the interplay and synergy between the human and biophysical subsystems as they are affected by processes operating at different scales. The framework argues for the need to include the following:

- the presence of perturbations and stressors, their sequencing, and interactions between them

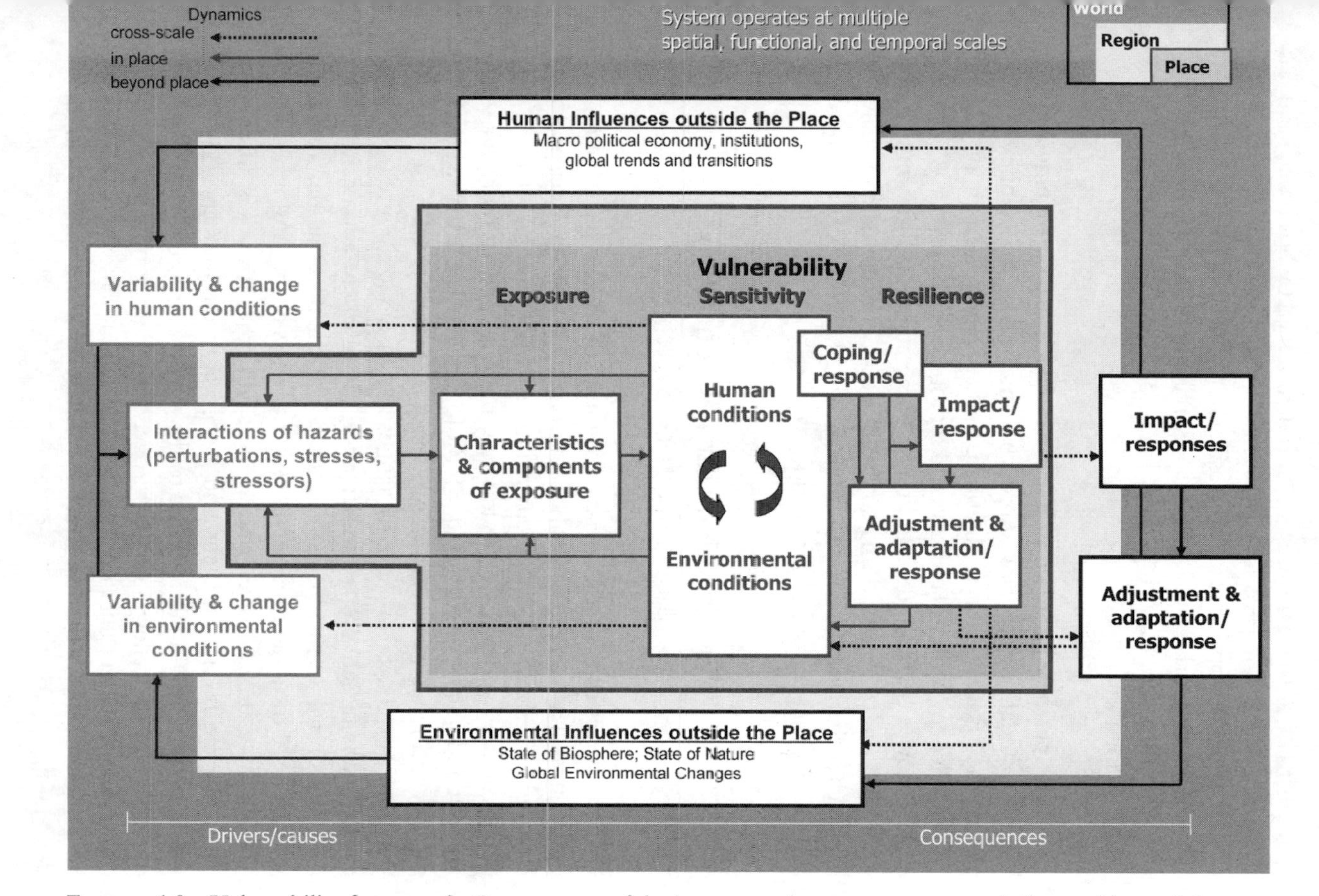

FIGURE 6.2 Vulnerability framework. Components of the human-environment system are influenced by and, in turn, influence factors outside the system of interest (Turner et al. 2003a). © 2011 by the National Academy of Sciences, USA.

- exposure to a perturbation and stressor, including the manner in which hazards are experienced
- sensitivity to the exposure, which is determined by both social and biophysical characteristics of a place
- capacities to cope or respond (resilience), including the consequences and attendant risks of slow (or poor) recovery
- the system's restructuring following responses taken (i.e., adjustments or adaptations)

Our experience in the Yaqui Valley helped lead to this framing and also laid the foundation for a new approach to measuring vulnerability of people and their places (Luers et al. 2003). In this chapter, we explore vulnerability challenges as they play out in this region of northwestern Mexico and explore how a better understanding of vulnerability can inform decision making in the region. In doing so, we illustrate conceptual and analytical approaches that could be useful in identifying and responding to vulnerabilities in other regions of the world.

Vulnerability in Water Resources and the Agricultural System

Other chapters in this book describe a range of shocks and changes that the Yaqui Valley has faced over the past decade. Socioeconomic shocks like a changing agricultural policy environment removed some financial support structures and exposed landowners and managers to more variations in input and output prices (chap. 8). Biological shocks such as whitefly invasion continuously force farmers to adapt their management and production patterns (chap. 9). Climate changes and drought are forcing farmers to consider alternate crops and irrigation management approaches (chap. 11). And on top of these changes themselves, the uncertainties associated with some of them have made planning and decision making very difficult. The ability of farmers to respond to these and other stresses is differential, depending on both their social and biophysical resources and how those resources are changing in response to the stresses. As a result, vulnerability of farmers in the valley is very heterogeneous, with some farmers experiencing more vulnerability than others. In the following sections, we provide a few examples that illustrate the different dimensions that underlie vulnerability.

Stressors in the Coupled Human-Environment System

The entire northwest coast of Mexico is arid. Winter monsoons trigger highly variable annual and decadal rainfall averaging 270 mm/yr. While the natural ecosystems of the valley are adapted to variable climatic conditions and occasional prolonged droughts, the agricultural system is far more sensitive to these variations. High winter rainfall lowers yields in winter wheat, perhaps because of lower than usual solar inputs, but also potentially due to interference with tillage and fertilizer application practices geared for dry soils. Prolonged droughts, such as the unusually extended period of low rainfall between 1997 and 2004, lead to dramatic declines in total reservoir volume. During that period, installed groundwater wells did not have sufficient capacity or quality to buffer reduced surface water storage, and thus deliveries of water to farmers declined from a long-term average of 2,655 million cubic meters (MCM) to 1,100–1,700 MCM over the 1997–2002 period (González et al. 1997; chap. 11). Not surprisingly, the Yaqui agricultural system is sensitive to climate variations.

Variation in water resources, of course, can have direct effects on irrigation and production, but the effects may also be indirect. For example, a high risk of salinization affects one-third of the valley's soils, those situated in the lower-lying portions of the valley and often associated with poor drainage and high water tables (chap. 9). Salinity levels threaten to reduce productivity over approximately 19,000 hectares. Management practices are in place to prevent much broader salinization problems (e.g., lowering groundwater tables, and the use of large amounts of relatively low-salt surface water in irrigation). These strategies rely on large supplies of freshwater. Drought response management strategies (e.g., deficit irrigation and substituting for surface water with more salty groundwater) may be suitable in the near term in many situations, but they can also cause significant problems in salinization-prone areas.

Conditions related to water sources and soil characteristics are only part of the story; policies can also impact different farmers in different ways. Policy reforms of the 1990s (chap. 8) reduced the government's involvement in agriculture, lowering credit subsidies, privatizing the fertilizer industry, decentralizing responsibility for irrigation to local user groups, and amending Article 27 to make it possible for *ejido* land to be legally rented and sold, among other things (see chap. 8; Naylor et al. 2001). These policy changes shifted more responsibility for agriculture from the government to farmers; they constitute a separate set of "stressors" in the valley that, like the biophysical stresses, are experienced differentially by different components of the human-environment system.

Application of the Vulnerability Framework: Responses to Drought

Given the range of socioeconomic and biophysical stresses that are experienced by the human-environment system of the Yaqui Valley, we evaluated some aspects of vulnerability and response in the valley, focusing on the water resources sector. Irrigation supports the large-scale agricultural economy of the arid valley; the eight-year drought of the late 1990s and early 2000s reduced reservoir levels and allocations to farmers. Overall, farmers adjusted to this stressor by reducing the number of irrigations and the area planted (primarily by eliminating summer production) and by experimenting with less water-intensive crops like sunflower and safflower (chap. 9). Some farmers have relied increasingly on groundwater pumping to augment decreased surface water access, although at very high energy costs (chap. 11).

In crisis years, the irrigation districts of the valley augmented their groundwater pumping capacity by developing new wells and increased pumping of saline groundwater. Too saline to apply directly to fields, most aquifer water is mixed with surface water until acceptable salinity levels are reached. This resource allowed the irrigation district to stretch its water supplies during the extended drought period (chap. 11). It also represented a valuable resource to the few farmers who owned private wells. The irrigation district acquired water from these (approximately fifty) private well owners by providing them access to additional deliveries of surface water for use or sale. Thus, producers with access to the natural capital of groundwater and the wells to access it were less vulnerable to the impacts of drought than those without.

Groundwater provided a supplementary water resource for those with reduced surface water access when sufficient surface water was available to dilute its salinity. Not all landowners who had access to groundwater could equally benefit from the more saline water, however. Nearly one-third of the valley soils tend toward high salinity and are thus more sensitive than other soils to salinization due to use of groundwater. These salinity-prone soils are more likely to suffer productivity declines during times of reduced surface water irrigation. Soil characteristics such as salinity and texture constitute a form of natural capital that influences farmers' vulnerability to drought.

With respect to water resource vulnerability, projections of anthropogenic climate changes in North America predict potential declines in precipitation for the Yaqui River watershed (USGCRP 2009). This is expected to result in increased drought and reduced surface water reliability (chap. 11). Climate change raises concerns about the long-term viability of today's large-scale irrigation practices, and thus the well-being and viability of the entire Yaqui Valley agricultural system. These changes in water resources are likely to play out differently for different farmers, based on their resource assets and abilities to respond to the changes.

Responses to Social Stressors

Most of our analysis focuses on water systems and drought; we also note that the effects of, and uncertainties related to, changes in markets and policy reforms—stressors on the social side—have been important in this human-environment system, and, as with the biophysical stressors, different farmers have different abilities to respond to them. Many of the agricultural reform policies put in place in the early 1990s substantially affected the institutions serving *ejido* farmers and landowners (see chap. 8 for a detailed analysis). Prior to 1990, *ejidos* accounted for approximately 56 percent of the total agricultural area and 72 percent of producers in the valley (CNA 1998; Puente-González 1999). Banking reforms affected the National Rural Credit Bank (BANRURAL) more significantly and negatively than other lending institutions, leading to reductions in the number of *ejido* farmers receiving credit and in total expenditures on subsidized interest. These reforms made it difficult for small, private, and *ejido* farmers to secure loans and access credit (Lewis 2002). Concurrent changes in land tenure policies (Article 27) allowed *ejido* land to be rented, sold, and mortgaged. By 1999, land area under *ejido* cultivation decreased dramatically, with 70 percent of interviewed *ejido* farmers renting their land, largely to private landholders (Lewis 2002). Almost all of these farmers reported difficult access to credit as a primary reason for renting. Farmers also noted increased prices for water and fertilizer inputs as impediments to farming (table 6.1). Yaqui Valley *ejido* farmers had difficulty adjusting to the policy reforms that exposed them to more variations in input and output prices. Many sold or rented their lands, and some landowners were able to earn more through land rental or sale than they did as farmers (Lewis 2002). The longer-term welfare and social impacts of land rental and sale are not known. Farm production has become more concentrated among fewer

TABLE 6.1 *Ejido* land rental and reasons cited for renting out.

Ejido *land rental*	*(%)*
Renting out	70
Renting in	4
Reasons cited for renting out	
Limited credit access	88
High agricultural input prices	67
High water price	63
Low crop prices	61
Lack of water	31
Lack of affordable machinery	24

Source: Lewis 2002; author survey of 70 *ejidatarios*, 1999. © 2011 by Wiley.

farmers, and the impacts of this on the human-environment system have not been assessed.

From Concepts to Measures of Vulnerability

Case studies like the one just discussed allow us to develop frameworks for analyzing vulnerability and increase our insights into why and how different human-environment systems are vulnerable (and what might be done to reduce vulnerability). We can build on these cases by developing vulnerability metrics that allow us to validate our hypotheses about the factors that underlie vulnerability and the interventions that can decrease vulnerability. Eventually, we can use vulnerability metrics to target interventions to vulnerable people or otherwise factor vulnerability concepts into policy and program design (Lobell et al. 2008).

An Assessment of Vulnerability of Wheat Yields

We estimated vulnerability in the valley by evaluating the susceptibility of selected outcome variables of concern (e.g., agricultural yield) to damage as a result of specific stresses (e.g., climate variability and change, and shifts in agriculture prices). We defined vulnerability as a function of the state of the variables of concern relative to a threshold of damage, the sensitivity of the variables to stresses, and the magnitude and frequency of stressors to which the system is exposed. Unlike many previous vulnerability metrics, our proposed metric is a discrete outcome measure (wheat yields) rather than a composite of proxy indicators (e.g., Moss et al. 2002; Kaly et al. 2002).

We used our metric to begin to evaluate the relative vulnerability of wheat yields to temperature variability and trends, and market fluctuations. Our wheat yield measures were based on yield estimates derived from Landsat TM and ETM+ data for four years—1994, 2000, 2001, and 2002—as described in detail by Lobell and Asner (2003). Building on the analysis of Lobell and Asner, we developed a linear least-squares regression model of yields accounting for average nighttime temperatures for January–April (Luers et al. 2003). We then calculated vulnerability using a threshold value of 4 t/ha, which is the approximate average minimum yield required for farmers to *break even* (i.e., zero net profit) based on the average management practices and commodity and input prices.

Our analysis suggests a skewed distribution of vulnerability to temperature variability and change within the study region, with most farmers exhibiting low vulnerability and a few farmers with high vulnerability. In addition, our method revealed that farmers' yields, absent any adaptation, are on average more vulnerable to a 10 percent decrease in wheat prices than a 1°C increase in average minimum temperature. Finally, we found that soils and management both contributed to relative vulnerabilities in the region, but that some of the constraints imposed by poor soil types can be overcome by improved management practices.

Developing vulnerability metrics for use here or elsewhere is complicated by a lack of baseline data, difficulty in measuring key social and cultural variables across scales and regions, and limitations in accounting for the dynamic character of vulnerability through time and space (Eakin and Luers 2006; O'Brien et al. 2004).

Vulnerability Analyses: The Way Forward

Our case study of vulnerability in Yaqui Valley highlights the need to address multiple and interacting hazards from the perspective of stakeholders who operate with different levels of natural and social capital. By focusing on the coupled human-environment system, the complexity of the system becomes apparent. We see that household sensitivity and responses to shocks depend in part on household access to biophysical resources—for example, soil type and salinity, water rights, land location—as well as to socioeconomic resources, like credit and technology, for example. Distinctions in access also lead to significant differences in individual household response options.

Identifying and better understanding vulnerability is just the first step. Vulnerability analyses can also help to identify critical adaptation needs and evaluate short- and long-term adaptation options (Adger, Dessai, et al. 2009; Adger, Eakin, and Winkels 2009; Moser et al. 2008; Moser and Luers 2008; Moser 2009). An understanding of who is vulnerable, why and how, can inform design and targeting of interventions to reduce vulnerability and facilitate adaptation to shocks. Very little research has been done on measures of vulnerability in human environment systems. Such research is important, and may be especially needed in countries and regions where populations' vulnerability is critically shaped by the intersection between exposure to stressors and structural inequalities (e.g., income, education, health, political power, etc.) (Lemos and Tompkins 2008). Such a research program, focused on integrated vulnerability and adaptation assessments has been called for in several recent reports (NRC 2009, 2010b).

Acknowledgments

We thank Bill Turner, Bill Clark, Robert Corell, Lindsey Christensen, Noelle Eckley, Grete Hovelsrud-Broda, Jeanne Kasperson, Roger Kasperson, Marybeth Martello, Svein Mathiesen, Rosamond Naylor, Colin Polsky, Alexander Pulsipher, Andrew Schiller, Henrik Selin, and Nicholas Tyler for their assistance in the development, analysis, and interpretation of this work in the context of our broader vulnerability research. Ivan Ortiz-Monasterio was there for every part of the work. This work was supported in part by National Science Foundation Grant BCS-0004236, with contributions from the National Oceanic and Atmospheric Administration's Office of Global Programs for the Research and Assessment Systems for Sustainability Program (http://sust.harvard.edu). It builds on collaborations sponsored by the Stockholm Environment Institute with Clark University, the Consortium for Social Science Associations, the International Human Dimensions Program, and the Land-Use/Cover Change Focus 1 Office at Indiana University.

Chapter 7

From Wheat to Waves and Back Again: Connections between the Yaqui Valley and the Gulf of California

MICHAEL BEMAN AND AMY LUERS

The Yaqui Valley is one of the most productive agricultural regions in Mexico, and the adjoining waters of the Gulf of California are home to one of the nation's most productive fishing industries. These two industries grew and prospered more or less independently for much their history. In fact, the Gulf of California was a source of food long before the Yaqui Valley was plowed and irrigated for wheat and other crops. Native peoples have lived along the shores of the gulf and fished its waters for millennia (Bowen 2004; Nabhan 2000; Robles-Ortíz and Manzo-Taylor 1972), while industrial shrimp trawlers began combing the gulf in the first few decades of the twentieth century (fig. 7.1). Then came agricultural development and the *green revolution*, which quickly transformed the Yaqui Valley from desert to wheat and created a new connection between land and sea—that of irrigation water and agricultural pollutants. The fishing and agricultural industries grew increasingly interconnected with the rise of the aquaculture industry in the late 1990s. Then, a series of privatization and liberalization policies combined with extended drought encouraged farmers to farm fish. As upland farmers turned to aquaculture, they were suddenly invested in the coast, and the terrestrial- and marine-based social and biological systems were connected to an unprecedented degree.

In this chapter we describe the gulf and coastal system at the edge of the Yaqui Valley; highlight some of the many connections between terrestrial and marine ecosystems and linked human systems of the gulf and

FIGURE 7.1 Fishing village Paredon Colorado, Yaqui Valley, circa 2000. (Photo courtesy of John Harrison.)

Yaqui Valley region; place these connections into the context of global trends; and consider future implications for the Yaqui Valley, Gulf of California, and other regions. In some ways, the Yaqui–gulf region functions as an analog for changes currently occurring throughout the world's ocean margins, yet the particular properties of the region also paint a picture of future changes that may be unique to the gulf and its adjacent lands.

Gulf of California Large Marine Ecosystem

The Gulf of California is a long (ca. 1,000 km), thin (ca. 100 km) body of water bounded on three sides by land—and by tall mountain ranges along the spine of Baja California, by the Sierra Madre Occidental of mainland Mexico, and by the desert (fig. 7.2). The gulf experiences limited freshwater inflow; large tidal cycles; hurricanes and other dramatic weather events; and its bathymetry is complex, consisting of a series of deep (up to 3,000 m) basins separated by shallow sills (500 m, in some places). All of these interact to produce a large marine ecosystem of enormous physical and biological complexity, which has supported a long history with an extensive fishing industry, and which continues to play a critical role in the region's

FIGURE 7.2 Sea-viewing Wide-Field-of-view Sensor (SeaWiFS) satellite image of the Gulf of California region taken March 22, 2001. The agricultural fields of the Yaqui Valley are visible along the east coast of the gulf, and the swirls in gulf waters are blooms of phytoplankton caught in oceanographic eddies.

economy through tourism, development, sport fishing, artisanal fishing, and industrial fishing.

At the base of this ecosystem are the microscopic plants of the sea known as phytoplankton, and even Spanish explorers recognized that uncommonly large numbers of these organisms are found in the Gulf of California: they reputedly referred to the gulf as the "Vermillion Sea" owing to the yellow coloring of its surface waters. Scientists from Scripps Institution of Oceanography in San Diego sailed to the gulf as early as the 1930s and measured primary production rates—essentially the bulk growth and activity of these organisms—using radioactive carbon-14 in the 1960s. Their production numbers from the gulf are among the highest found in marine waters worldwide and are comparable to other rich parts of the sea (Zeitschel 1969).

However, conceptualizing the Gulf of California (or Sea of Cortez) simply as a productive body of water that supports fisheries is less than half the story. The region is designated by UNESCO as a World Heritage Site due to the fact that "almost all major oceanographic processes occurring in the planet's oceans are present," along with 891 species of fish, 39 percent of the world's marine mammal species, and fully one-third of all cetacean species on Earth (http://whc.unesco.org/en/list/1182/). Hence, many environmental nongovernmental organizations, including the World Wildlife Fund, The Nature Conservancy, and the Environmental Defense Fund, are all actively working to preserve critical portions of this unique body of water that faces many stresses.

No area of the surface ocean is untouched by human activities (Halpern et al. 2008). Even in the middle of the seas, temperatures tick upward as a consequence of climate change (IPCC 2007b), aerosol deposition alters seawater chemistry (Duce et al. 2008; Paytan et al. 2009), and *ocean acidification*—produced by dissolution of carbon dioxide in seawater—reduces ocean pH (Sabine et al. 2004). As these physical and chemical changes pervade the deep ocean, exploitation of global fisheries (Myers and Worm 2003; Worm et al. 2006), expansion of aquaculture (Naylor et al. 2009), changes in the chemistry of river flows to the sea (Rabalais et al. 2002), and appropriation of freshwater for human use (Postel et al. 1996) are further transforming marine ecosystems. On a global map of human impacts on marine ecosystems, 41 percent of the ocean is strongly affected by multiple anthropogenic drivers of ecological change. The Gulf of California experiences medium to medium-high impact (Halpern et al. 2008). Changes in the Gulf of California and the surrounding land, islands, and the Yaqui Valley fit within a larger picture of global environmental change—what is occurring in the Gulf of California is not unique. What is unique about the Gulf of California and the Yaqui Valley is how tightly coupled different parts of the gulf are, and how rapidly environmental transformations are taking place in a region that is still sparsely populated. The pace and extent of these changes holds lessons for other regions of the ocean, and more fully understanding them will be critical for maintaining the ecological integrity of the gulf and the resources it provides.

Sea to Land and Land to Sea Connections in the Gulf of California and Yaqui Valley Region

The connections that link land and sea in this region are strong and have been both strengthened and stretched, concurrently and in multiple ways, over the past century due to the influence of human decisions and actions both on land and on sea. In this section, we outline some of the most important biophysical and social mechanisms that have influenced this integrated land-sea system, and some of the most important changes that have occurred to affect them.

Gulf of California Seascape and Connections to Land

The Gulf of California is formed by the same forces that shake buildings in San Francisco and Los Angeles; the movement of the Pacific plate relative to the North American continental plate splits and widens the gulf over time. Just a few hundred miles from the Yaqui Valley, under several thousand meters of water, some of the most well studied deep-sea vents in the world occur at the bottom of the Guaymas basin as a result of this rifting. The unique shape of the gulf is produced by these geological forces and sets the stage for much that subsequently occurs within the gulf ecologically. In particular, mountain ranges along both Baja California and mainland Mexico effectively "polarize" winds, such that they travel essentially north or south across the gulf (fig. 7.2). Coupled with the heating and cooling of the Colorado plateau, the near-monsoonal winds of the Gulf of California reverse from a northwesterly direction during winter to a southerly or southeasterly direction during summer (Pares-Sierra et al. 2003). Northerly winds drive intense "upwelling" along the eastern gulf in winter—as they do in upwelling systems throughout California and other portions of the Eastern Pacific Ocean. (During upwelling, winds push surface waters offshore, and this water is replaced by colder, nutrient-rich water that is upwelled from below.) As the winds reverse course in summer, weak upwelling occurs in the west gulf along the coast of Baja. Upwelling is weaker during summer because the gulf heats up rapidly—water temperatures can reach nearly 90°F (Beman et al. 2005)—and becomes highly stratified (White et al. 2007). In this way, the gulf is really two gulfs that vary with the seasons: one of active upwelling during winter and one of intense stratification during summer.

This seasonal oceanography and meteorology has a number of implications for the Yaqui Valley. Onshore monsoon winds in summer bring moisture and rain with them, and in fact most precipitation in the valley and the watershed that feeds its reservoirs falls in the summer (Nicholas and Battisti 2008). Commencement of upwelling in November and its cessation in May also neatly match the typical cropping cycle in the Yaqui Valley. Interestingly, Lobell and Ortiz-Monasterio (2007) found that wheat yields were often related to the minimum temperature in January, and that this in turn may be related to minimum sea surface temperatures (SSTs; D. Lobell, personal communication). In this way, the larger climate system of the valley region—controlled in part by oceanography—ties land to sea.

A stronger connection results from the biological effects of upwelling: as nutrients are brought up into the light, microscopic phytoplankton rapidly grow and photosynthesize, and this results in pulses of biological production that percolate through the ecosystem. Different regions of the gulf behave in different ways oceanographically, resulting in different phytoplankton provinces and patterns of productivity, whether measured by hand decades ago (Gilbert and Allen 1943; Round 1967) or remotely sensed by satellites (Santamaría del Angel et al. 1994a; Kahru et al. 2004). These patterns in turn affect the distribution of larger organisms that feed on phytoplankton (Brinton and Townsend 1980), and eventually fish and mammal distributions (Aceves-Medina et al. 2004).

For many islands of the gulf, the productive sea is a major source of food and nutrients for land ecosystems (Polis and Hurd 1996; Polis et al. 1997; Anderson and Polis 1998; Stapp et al. 1999). These are provided via two routes—marine debris that wash up on land, and seabirds that prey on marine organisms and roost on land (Polis and Hurd 1996; Stapp et al. 1999)—and impart the unique isotopic signature of marine ecosystems upon terrestrial ecosystems. Globally, there is a growing appreciation of subsidies between sea and land, including salmon-derived nutrients in the Pacific Northwest (Holtgrieve et al. 2009) and the global transport of marine nutrients to land from fisheries (Maranger et al. 2008). In the Gulf of California, the juxtaposition between desert and the rich sea is most striking on islands, and the important work of Gary Polis, who tragically passed away while conducting fieldwork in the gulf, has clearly documented important sea-land ties. Marine nutrient subsidies to mainland ecosystems have not been quantified, but are likely less important than on islands; Polis and Hurd (1996) suggest, however, that proximity between productive marine ecosystems and relatively depauperate ecosystems has the potential to produce similar patterns and flows elsewhere, and rich upwelling areas can be found near many deserts. In the Gulf of California, any disruption of marine ecosystem processes has the potential to strongly affect island ecosystems—many of which are protected, owing to unique endemic species.

Changing Landscapes and Connections with the Sea

Several changes are clearly evident in the gulf as a result of activities on land, and one of the most profound has been the damming and diversion of the Colorado River in the northern gulf. Little or no water now enters the gulf in the Colorado River estuary, as this water is used for drinking water and farming throughout California, Arizona, Utah, Nevada, and Colorado. The chemical composition of water that does reach the gulf is radically different, with higher salinities, higher levels of toxic metals and chemicals (García-Hernández et al. 2001), and higher nutrient concentrations (Hernández-Ayón et al. 1993). There is evidence that heavily modified flows also reduce shrimp landings in the northern gulf (Galindo-Bect et al. 2000)—indicating that heavy modification of the Colorado River can affect the gulf ecosystem in detectable ways.

In our own work in the Yaqui Valley, farther to the south, we documented large phytoplankton blooms in the Gulf of California that closely tracked valleywide irrigation events (chap. 10; Beman et al. 2005). These events are a consequence of nitrogen-rich runoff from the valley entering gulf waters, which are critically depleted in nitrogen compared with concentrations of other required nutrients. This depletion occurs both in surface waters and nutrient-rich upwelled waters, so the natural productivity of the Gulf of California may be augmented by nitrogen runoff from the fields of the Yaqui Valley. The downstream effects of these blooms are not yet known, and because the gulf ecosystem is naturally productive, it is potentially adapted and equipped to effectively process such events. On the other hand, it could be the case here, as in many other regions, that such blooms are harmful to marine life, or that oxygen is depleted as bloom material sinks through the water column, or that bloom effects are coupled with other changes in the gulf. What is clear is that any possible effects are not localized; due to the unique oceanography of the Gulf of California, large eddies rapidly transport water masses across the width of the gulf (Pegau et al. 2002), and blooms off the coast of the Yaqui Valley can reach Baja California in a matter of days. These circulation patterns tightly connect waters throughout the gulf, and so changes along one coast may impinge upon distant shores (fig. 7.2).

Changes along the Shore

In 2004, a group of scientists and writers traveled along much of the Gulf of California shoreline tracing the path of the marine biologist Ed Ricketts and the writer John Steinbeck, who explored the gulf in 1940 and documented their findings in *The Log from the Sea of Cortez*. Sagarin and colleagues uncovered significant changes in the intertidal communities of the gulf, and "although [they] found many of the same species as [Steinbeck and Ricketts] did, populations were, in many cases, not as geographically widespread, and individuals were fewer in number and smaller in size" (Sagarin et al. 2008). The most dramatic changes occurred in Cabo San Lucas, which has boomed from a small village to a major tourist destination. With changes in land tenure laws in the 1990s (Luers et al. 2006; chap. 8), large portions of the gulf region now face the prospect of similar forms of development for obvious reasons: a sunny climate, beautiful beaches, and a stunning backdrop of desert and sea. An *Escalera Nautica* or Nautical Staircase of multiple marinas has also been proposed for the gulf to encourage tourism. Much of the Gulf of California shoreline will always remain undeveloped given its topography and aridity; however, explosive growth in Cabo San Lucas, Loreto, San Carlos, and Puerto Peñasco may be a harbinger.

Sagarin and colleagues also noted a sharp change in the pelagic ecology of the gulf, particularly in the rise of the "jumbo" or Humboldt squid, *Doscidicus gigas*. These voracious predators eat a range of organisms in the gulf (Markaida and Sosa-Nishizaki 2003)—including each other—and grow a thousandfold in one to two years (Nigmatullin et al. 2001; Markaida et al. 2004). The Gulf of California fishery for these squid is now one of the largest squid fisheries in the world, removing approximately 120,000 tons of biomass a year (Gilly et al. 2006), yet this fishery was nonexistent before the 1970s (Rosas-Luis et al. 2008). The rise of a newly important predator and emergent fishery is part of a story of long-term change in the marine fisheries of the gulf. Evidence drawn from accounts of early travelers (Sáenz-Arroyo et al. 2006), shifting baselines among three generations of fishermen (Sáenz-Arroyo et al. 2005), compilations of interviews and statistics in the northern gulf (Lozano-Montes et al. 2008), and complications of interviews, statistics, and field surveys in the southern gulf (Sala et al. 2004) all point to the same conclusion: fishing and other perturbations have transformed the ecology of the gulf. Fewer species, less desirable species, and smaller individuals are caught—with more effort—as time passes.

Blurring the Line between Land and Sea

As many fisheries have declined, aquaculture has boomed over the last decade, transforming the coastal landscape (including large portions along the edges of the Yaqui Valley) from desert brush, wetlands, and mangroves to shrimp ponds (fig. 7.3; Luers et al. 2006). From 1990 to 2001, the shrimp farm area in southern Sonora grew almost 12,000 hectares (see fig. 4.3 in chap. 4).

Environmental consequences have resulted from the discharge of aquaculture waste into the gulf (Páez-Osuna et al. 2003), as well as the loss of mangrove ecosystems. Mangroves are remarkable salt-tolerant plants found in tropical and subtropical coastal regions worldwide, and mangrove forests are renowned for their provision of valuable ecosystem services (Feller et al. 2010). In the Gulf of California, Aburto-Oropeza et al. (2008) found a strong correlation between mangrove habitat area and fishery yields—in terms of both biomass and monetary value. They calculated that annually mangroves provide approximately US$37,500 of mangrove-related fish and crab per hectare. Because the size of this resource varies from year to year but is continuously renewable, mangroves become enormously

FIGURE 7.3 Aerial photo of shrimp ponds in the Yaqui Valley.

valuable over time—five hectares produce nearly US$1 million worth of fish and crab in just six years (Aburto-Oropeza et al. 2008). Given the values of these artisanal fisheries, the destruction of mangroves for shrimp aquaculture appear to make little economic sense given aquaculture's lack of profitably, the common abandonment of ponds, and the slow return to mangrove forest (Luers et al. 2006).

The rapid growth in the aquaculture industry has also contributed to dramatic changes in the flow of goods and services from the uplands to the coasts, and vice versa. The results of our study suggest that the privatization and liberalization reforms promulgated in the 1990s influenced the shifts in land tenure and land use along the coastal zone of southern Sonora, facilitated the rapid growth of the shrimp farming industry, and forged new linkages between the coastal and upland human-environmental systems (see chap. 8). The policy reforms of the 1990s provided the legal basis for the division of communal lands and their transfer from the *ejido* to the private sector. Prior to the reforms, most coastal lands were *ejido* lands, which were held collectively with few illegal rentals, and shrimp farms were predominantly built on non-*ejido* lands. After the reforms, coastal lands were parceled out and rapidly changed hands as private shrimp farm development spread into former *ejido* lands. Because a number of these new private aquaculture farmers also had investments in the upland agricultural sector, this shift tightened the links between the coast and the uplands.

Meanwhile the *ejido* shrimp farms that did operate often had poor financial performance. Our research indicates that between 1998 and 2001, most *ejido*-run farms provided little or no profit, with many going into debt (Luers et al. 2006). Among the *ejido* farms there was a large variability in annual profits due to disease problems, natural disasters, and differences in management approaches. In 2000 and 2001, at least two-thirds of the *ejido* shrimp farms in the region had outbreaks of viruses, including white spot and taura viruses that reduced shrimp quality and yields, and produced no profit distributions to the *ejidatarios*. This resulted in many coastal *ejido* workers once again turning upland to seek employment.

Perspectives on Connectivity and Change in the Gulf of California and Yaqui Valley

Many of the changes documented in the Gulf of California represent fundamental physical or chemical transformations that are not easily undone. The damming of the Colorado River, agriculture in the Yaqui Valley, and coastal aquaculture all amend and modify the chemical composition of freshwater and seawater (Hernández-Ayón et al. 1993; García-Hernández et al. 2001; Harrison and Matson 2003; Páez-Osuna et al. 2003; Harrison et al. 2005; Beman et al. 2005). Changes brought on by global warming and ocean acidification may also have important ramifications in the gulf, where physical oceanography is tightly coupled to biological oceanography and strongly connected to marine ecology. For example, El Niño-Southern Oscillation (ENSO) events can have drastic effects in the Gulf of California, with increased rainfall leading to much higher terrestrial primary production (Polis et al. 1997) that contrasts sharply with steep declines in marine primary production (Santamaría del Angel et al. 1994b; Kahru et al. 2004). The enormous amounts of energy, water, and biomass that flow through the gulf can therefore be altered by climate-driven changes in ocean circulation.

Perhaps the most significant change that could occur in the gulf is the spread of a naturally occurring oxygen minimum zone. Low oxygen water is found throughout the eastern Pacific Ocean and extends into the Gulf of California, where waters from ~250 meters to 800 meters are heavily depleted in oxygen. A 2008 paper in *Science* presented evidence that these regions are expanding laterally and vertically across the globe—including portions of the eastern Pacific (Stramma et al. 2008). Oxygen declines are evident off the coast of Southern California (Bograd et al. 2008), and a low-oxygen "dead zone" has developed off the coast of Oregon as a possible consequence of climate change (Chan et al. 2008; Grantham et al. 2004). Hard-and-fast evidence for expanding oxygen depletion in the gulf is not yet available, but this is of obvious consequence given the importance of oxygen for many marine organisms, from fish to marine mammals. Any efforts to protect endangered marine species, or to protect unique ecosystems such as the islands of the gulf, must contend with these and other chemical and physical shifts.

Coupled with massive fishing pressure, habitat destruction, temperature increases, oxygen declines, and ocean acidification, the elevated concentrations of nutrients, trace metals, and artificial chemicals stretch the strong ties between land and sea and sea and land in the Gulf of California. Over time, they threaten the ecological integrity and viability of the Gulf of California. However, interdisciplinary research conducted in the gulf region by our group indicates that the problems of agricultural runoff are solvable. Embracing site-specific, efficient management approaches in the Yaqui Valley can be a win-win for agriculture and the environment in the region, and many farmers appear to be awakening to this possibility (chaps. 3 and 9). Current practices in shrimp aquaculture could also be replaced by more sustainable approaches.[1] Efforts by NGOs and others in the region[2] to develop and implement these approaches are ongoing. In addition, strengthening of the general trend toward sustainable seafood consumption[3] may be required. Solutions to these and other problems have evolved and will continue to evolve; understanding the Gulf of California as a dynamic, complex, large marine ecosystem that is closely coupled with actions on land is key to the development of solutions. Settlement and development of the Gulf of California region—from native peoples to Cabo San Lucas—has centered on the sea; we suggest that the future of the region will be strongly tied to the sustainable management of the Gulf of California, its adjacent landscapes, and their unique resources.

Acknowledgments

We thank Bill Gilly for many insightful discussions in California and in the Gulf of California; Esther Cruz-Colin and Juan Delgado for fruitful collaboration and many good laughs; Ivan Ortiz-Monasterio for all of his help in the field and at CIMMYT; Peter Jewett and Eve Hinckley for assistance in the field and lab; and John Harrison for a great introduction to the Yaqui Valley. Thanks to Pam, Roz, Wally, David, Toby, Lee, Ellen, and all of the Yaqui project. Thanks are also due to current and recent collaborators in the Gulf of California, including Brian Popp, Fred Prahl, Jason Smith, Angel White, Rachel Foster, Jackie Mueller, Natalie Walsgrove, Julie Stewart, and Jorge Ramos. Funding was provided by the US National Science Foundation Chemical Oceanography program (OCE-0824997 and OCE-1034943) and Office of International Science and Engineering (OISE-0438396), as well as by the MacArthur Foundation, Packard Foundation, Switzer Family Foundation, and the Teresa Heinz Environmental Scholars Program.

PART III

Elements of the Yaqui Valley System

Chapter 8

The Yaqui Valley's Agricultural Transition to a More Open Economy

ROSAMOND NAYLOR AND WALTER FALCON

Assessing agricultural practices and their environmental impacts at the regional scale is often problematic for policy analysts. And so it is with the Yaqui Valley. The main economic and policy determinants of change for farmers have originated at national and international scales, and have mostly been exogenous to production systems in the region. Since the start of our study in the early 1990s, farmers in the Yaqui Valley have had to respond to fluctuating prices for agricultural outputs and inputs caused in part by the North American Free Trade Agreement (NAFTA);[1] a set of national liberalization policies for agriculture implemented in Mexico City; a constitutional change in the rules governing collective (*ejido*) land holdings throughout the country; macroeconomic policy shifts related to foreign direct investment and current account imbalances at the federal level; and swings in international commodity prices driven, for example, by changes in China's production and consumption.

Our analytical task would have been made easier if the world had simply stood still for a decade. Unfortunately, international capital and

---

The field-based policy work reported in this chapter was largely completed in 2007. Since that time, world commodity prices have changed substantially. A new Stanford survey is underway that will reexamine private and social profitability of Yaqui farming systems, as of 2011. Our expectation is that a supplemental article on "Yaqui Revisited" will be available in late 2012.

commodity markets never rest, and policies constantly change at the national scale, often irrespective of their impacts on agriculture, the environment, or specific populations in the rural economy. The purpose of this chapter, therefore, is to analyze how the complex web of macroeconomic, trade, and agricultural sector policies—coupled with drought, pathogen, and commodity-market shocks—have influenced agricultural practices and economic returns in the Yaqui Valley. The analysis centers on a sixteen-year period from 1991 to 2007 when commodity and financial markets demonstrated large variation.

A broad focus on policy in the context of agricultural change in the Yaqui Valley raises several more general methodological points. First, suggestions for new policies should not necessarily be viewed as a "fix" for environmental and equity problems in a region's agriculture. The historical development of one policy overlaid by another is part of the underlying cause of social problems within agriculture in many countries, and initial policy advising is often best targeted at dismantling existing policies rather than implementing new ones. A second and related point is that policies focused on income and employment generation in developing countries like Mexico—and even industrial countries like the United States—often override environmental policies in agriculture. By the same token, macroeconomic and trade policies, which are often geared toward growth in the manufacturing sector, have a tendency to override policies that might better serve the agricultural sector. Finally, the effects of policy on rural development in the Yaqui Valley are not representative of Mexico as a whole. Mexico is known for its dual agricultural sector, with high-input commercial systems characteristic of Sonora on one end of the spectrum, and low-input subsistence systems, more pronounced in the southern maize-growing states, at the other end. Nevertheless, the Yaqui Valley, as home to the *green revolution* in wheat and still one of the more productive wheat systems in the world, serves globally as an important model system. In particular, the conclusions that we draw for the Yaqui Valley about the importance of economic and biophysical shocks, the difficulties of agricultural diversification, and the increasing dilemmas faced by very small farmers should resonate in progressive regions throughout the world.

With this perspective in mind, the goals of this chapter are to review the suite of policies that influence agricultural systems and to illustrate how the effects of policy on rural society might be assessed, using the Yaqui Valley as a study site. Disentangling the effects of planned agricultural and trade reforms, radical shifts in macroeconomic policy, and unexpected biophysical shocks is virtually impossible. The chapter therefore provides a brief explanation of the major changes; describes some of the salient farm-level and valleywide responses to these changes; and then uses comparisons of financial and economic profitability throughout the fifteen-year period to summarize some of the larger efficiency effects.[2] The profitability analysis is particularly important for gauging economic sustainability of crop production in the Yaqui Valley and the overall costs to society for subsidizing particular crops, such as wheat, and particular farm inputs, such as water. Qualitative evidence is also presented on some of the important environmental, health, and equity consequences of policy and institutional change in the valley, all of which are challenging to quantify in terms of monetary costs and benefits. The chapter ends with a summary of what has been learned—and what research opportunities might have been missed—in the policy analysis process during the course of this study.

Agricultural Reform and the Transition to Globalization

The history of the Yaqui Valley, described in chapter 2, demonstrates a period of marked changes during the past two decades. Farmers faced drought, pest, and disease constraints on their traditional cropping system, and unprecedented exposure to international markets and fluctuating macroeconomic conditions. Here we focus specifically on policy-induced change in the valley, drawing heavily on earlier project papers by Naylor, Falcon, and Puente-González (2001); Puente-González (1999); Lewis (2002); Luers (2004); Kitzes (2005); and Luers, Naylor, and Matson (2006). Our analysis is anchored with information from a series of field surveys in the Yaqui Valley conducted by Stanford University and the International Maize and Wheat Improvement Center (CIMMYT).[3] The chapter also contains data and literature through 2007 to pave the way for a more forward-looking discussion on farm practices in the region.

During the 1980s, government involvement in almost all phases of the Mexican food system was pronounced. Significant price supports for agricultural products, large input subsidies on water, credit, and fertilizer, and major consumption subsidies on basic food products were justified primarily as poverty-alleviation policies. Although Mexico became a full member of the General Agreement on Tariffs and Trade (GATT) in 1986 (an international institution that evolved into the World Trade Organization [WTO] in 1995), few significant reforms were undertaken in the agricultural sector at that time.

By 1990, government expenditures on agriculture were so complicated in Mexico that it was virtually impossible to assess the net effect of the policies on efficiency, equity, or the environment. What was clear, however, was that the expenditures were large in absolute terms—the equivalent of US$6 billion[4] or about 13 percent of the Mexican federal budget in 1990 (OECD 1997). The policies also severely distorted relative prices within agriculture. Overall, the producer subsidy equivalent (PSE) for agriculture—the share of farmers' revenue derived from subsidies—was about 30 percent of the value of agricultural production in 1990. The consumer subsidy equivalent (CSE) for food products was about minus 5 percent—the negative estimate implying that food consumption was being subsidized on a net basis relative to a world price standard (OECD 1997).[5]

In the early 1990s the Salinas administration began to withdraw government support in agriculture as part of a broader liberalization process that was occurring in other sectors of the economy. Agricultural reforms throughout the decade were impressive in both scope and scale, and they replaced most of the policies of the 1980s. The five key components included the following:

1. Development of a fifteen-year program of direct income payments to farmers (PROCAMPO), which was linked to the abolition or reduction of a wide array of input subsidies and price supports that were coupled to production
2. Installation of new international trading arrangements for agriculture, mainly via NAFTA, which reduced trade barriers, motivated large changes in prices of many agricultural inputs and outputs, and thus dramatically altered relative prices to producers
3. Reduction of the government's institutional involvement in agriculture, such as downsizing the National Basic Commodities Company (CONASUPO), privatizing the Mexican Fertilizer Company (FERTIMEX), removing or reducing government credit subsidies (BANRURAL), and largely eliminating public extension services
4. Decentralization of operating authority and funding responsibilities for irrigation systems to local water-user groups via the Water Laws of 1992 and 1994
5. Amendment of Article 27 of the constitution of Mexico, which made possible the (legal) sale and rental of *ejido* land

The overall effect of these reforms shifted relative responsibility for agriculture from the government to the private and *ejido* sectors, from federal authority to regional responsibility, and from government-determined pricing rules to the marketplace. Several of these reforms had antecedents in earlier periods, but what is remarkable about this set of policies is that they largely came into existence within the three-year period from 1992 to 1994. The reforms also coincided with a major macroeconomic shock driven by a move from a fixed to a floating exchange rate, and a set of biological shocks (described in chap. 2) that altered traditional cropping patterns in the valley. Not surprisingly, many rural communities were ill prepared for the rapid changes that followed. They found themselves in rather desperate searches for new sources of income and new farming methods that could reduce costs and increase revenues.

An important lesson from this period is the difficulty in analyzing the effects of one policy, such as NAFTA, "holding all else constant."[6] Here we summarize the key changes in agricultural and trade policy that affected Yaqui producers in the 1990s and some of their responses to these changes at the farm and valley scale. Additional responses, especially those involving fertilizer and irrigation water, are described in chapters 9 and 11.

Toward Decoupled Policies

Policies adopted in the 1990s were aimed at increasing the efficiency of the agricultural sector by opening it more to international competition and by reducing the extent of government intervention in marketing. Perhaps the most important policy change was the movement away from direct price supports for staple grains, beans, and oilseeds—administered through CONASUPO—to direct income payments to producers through the PROCAMPO program. The shift toward decoupled support of the agricultural sector coincided with Mexico's entrance into NAFTA in 1994. It also preceded a broader global mandate to eliminate distortions caused by agricultural price policies undertaken as part of the WTO in 1995.

Prior to the reforms, CONASUPO provided price supports to domestic producers of staple grains, beans, and oilseeds (Yúnez-Naude 1998). Like many state trading enterprises in developing countries, CONASUPO bought commodities at prices above world prices and then subsidized sales to consumers, thereby insulating Mexico from international competition. In addition, CONASUPO maintained control over imports and exports and had subsidiaries involved in storage, processing, marketing, and distribution. Parallel marketing institutions also existed for other crops, such as cotton, coffee, and sugar.[7] In the case of wheat, the most important crop in the Yaqui Valley, CONASUPO sold domestic purchases and imports of bread wheat to millers at a price equivalent to or below the purchase price. Farmers in the high-yield, price-supported environment of the Yaqui Valley were thus able to secure relatively large and stable incomes on a systematic basis.

The restructuring of CONASUPO began in the 1980s and continued through the early 1990s (King 2006). By 1991, ASERCA (Support and Services for Agricultural Marketing) within the Ministry of Agriculture had oversight responsibilities for marketing all staple crops, apart from maize and beans. The purchase and storage of agricultural commodities were placed in the hands of the private sector. The main responsibilities of ASERCA were to facilitate market development for wheat, sorghum, rice, soybeans and other oilseeds and to promote the export of cotton, fruits, and vegetables. Unlike CONASUPO, ASERCA did not purchase agricultural commodities from producers, but instead provided market information and helped to establish regional and international distribution channels.[8] CONASUPO was formally abolished in May 1999 by presidential decree (OECD 2000).

A fifteen-year program of direct income payments was introduced through PROCAMPO in 1994. Payments were based on the historical area of farmland devoted to wheat, maize, beans, cotton, rice, soybeans, safflower, sorghum, and barley; they were not related to the level of current output (hence, "decoupled"). The idea was to provide a greater role for markets in determining production decisions based on available inputs for each region. In practice, larger farmers continued to receive the greatest share of support. However, PROCAMPO also benefited some subsistence farmers who previously did not receive price supports, and to a lesser extent it helped farmers on small, low-yielding plots where the value of price supports was minimal.

The transition to decoupled policies had implications for the national agricultural budget, as well as for the incomes of farmers in the Yaqui Valley. In 1991, 15 percent of the agricultural budget was spent on price supports and subsidies; in 1996, only 7 percent of the agricultural budget was spent on these programs, and almost 30 percent was spent on direct income payments (OECD 1997). Moreover, the real value of the agricultural budget in 1996 was only about half of its value in 1991. As a combined consequence of changed international prices, macropolicy, and agricultural policy, the PSE (producer subsidy) for aggregate Mexican agriculture fell from 26 percent to 5 percent between 1991 and 1996. During the same period, the CSE (consumer subsidy) also went from moderate subsidization (–22 percent) to low taxation (+6 percent) relative to international price norms (OECD 2007).

Farmers in highly productive areas like the Yaqui Valley were the biggest losers (in absolute value) from the policy shift. Although farmers benefited some from income payments, PROCAMPO payments by design represented only a small share of total revenues—typically about 5 percent. The PROCAMPO payments thus provided only meager compensation for the removal of the price supports that had prevailed earlier in high-yielding areas.

Toward Decentralized Factor Markets

During the 1980s the government launched a broad effort to privatize or eliminate public enterprises across all sectors of the economy and succeeded in reducing the total number of state-run firms, from 1,155 in 1982 to 280 in 1990 (King 2006). This process continued into the early 1990s, when government-held companies that produced fertilizers, seeds, and other inputs were privatized. In addition, the credit, land, and water markets were privatized, and public extension services were largely dismantled with the expectation that private delivery would replace them. These changes affected input prices, competitiveness, and farm practices in ways intended to push farmers toward cropping systems that adhered more closely to Mexico's international comparative advantage—particularly in export-oriented regions like the Yaqui Valley.

Fertilizer

During the 1980s, domestic fertilizer prices were held below international prices through a series of budget transfers to the state-owned fertilizer company, FERTIMEX. In addition, energy inputs and ammonia used to produce fertilizers were subsidized through the Mexican Petroleum Company, PEMEX. Direct subsidies on fertilizer were eliminated with the privatization of FERTIMEX in 1992, when it withdrew from retail distribution. At the same time, indirect subsidies through PEMEX were reduced, although some preferential pricing for ammonia to private companies persisted. The direct subsidy on urea (comparing domestic and international prices) fell from 24 percent in 1989–91 to 0 percent in 1993–96. In 1996, real urea and triple super phosphate prices were 85 percent and 53 percent higher, respectively, than 1988 prices for the average Mexican farmer (Puente-González 1999).

Credit

Mexico's financial system also went through a major reform in the late 1980s, which led to a reprivatization of commercial banks in 1992 (OECD 1997; Nadal 2003). Farmers in Mexico have traditionally been credit constrained. Relatively high risks associated with agricultural conditions (especially in rain-fed areas), high and volatile interest rates associated with rapid rates of inflation and changes in macroeconomic policies, and lack of collateral (particularly for *ejidatarios*), have created the largest barriers to lending. To compensate for these difficulties, the Mexican government had earlier supported agricultural lending through the promotion of public-sector development banks (as discussed in chap. 2). In 1988, roughly 80 percent of total agricultural lending in Mexico was done through these development banks, and interest rate concessions accounted for over one-third of total government expenditures on agriculture (Puente-González 1999; OECD 1997).[9] Loans were made to producers at subsidized (and sometimes zero) rates without collateral, and the government covered the operating deficit of the development institutions. Moreover, loan repayments by many farmers—even at the subsidized rates—tended to lag by months and even years.

The restructuring of banking that began in 1989 occurred mainly within BANRURAL (the National Rural Credit Bank), the primary bank servicing the *ejidos*. By 1996, the area covered by BANRURAL loans dropped from seven million to less than two million hectares, and the number of producers receiving credit dropped from 1 million to 500,000 (Puente-González 1999). As the role of BANRURAL declined, other institutions assumed some lending responsibilities.[10] Nevertheless, total expenditures on subsidized interest fell by over 75 percent, making it much more difficult for small private farmers and *edijatarios* to secure loans.

Land

The change in credit conditions occurred simultaneously with an important change in land policy within Mexico that is described in chapter 2. The amendment to Article 27 of the 1917 Mexican Constitution (the Agrarian Law of 1992) altered over seventy years of land policy with respect to *ejido* land ownership (Lewis 2002). The stated goal of the policy was to modernize the agricultural sector, thereby appeasing critics of the *ejido* system within the Salinas administration who believed that inefficiencies associated with collective land tenure and small plot sizes impeded agricultural growth (Gates 1993). The reform of Article 27 set up a flexible land-tenure regime inside *ejidos* by allowing communal and individual property holders to rent, sell, and mortgage land legally for the first time. New institutions, such as PROCEDE (Program for the Certification of *Ejido* Land Rights and the Titling of Urban House Plots), were established to facilitate this process and to certify official land titles (Lewis 2002; Luers et al. 2006). Joint ventures between the *ejido* sector and the private sector were also encouraged. A legal market for *ejido* land was thus established, replacing the illegal market that was widely acknowledged to exist. When coupled with changes in government credit and water policies, the reform of Article 27 accelerated increases in private ownership and in the operational size of farms within the valley.

This policy also had important implications for coastal *ejido* communities in the Yaqui Valley (Luers et al. 2006). Until the mid-1970s, most of the coastal lands in the region were federal territory, although some *ejido* communities—predominantly fishing communities—had a long history of living on the land. The first large distribution of coastal lands to *ejidos* took place in the late 1970s under the Echevarría administration; at the time, these lands seemingly offered little productive potential. The second major coastal land distribution was carried out by the Salinas administration in the late 1980s as part of the Integral Agrarian Program of Sonora (PAIS) and established the first aquaculture *ejidos* in the region. By the early 1990s, the view of coastal lands in southern Sonora had substantially changed from being a "wasteland" to being a potential "goldmine" for shrimp aquaculture (chap. 7). The introduction of Article 27 opened these public coastal goldmines to private sales and rentals.

Water

Rural areas were also significantly affected by the decentralization of control over irrigation water, which is described in greater detail in chapter 11. The transfer from the federal government to local user groups originated largely in Mexico's 1982 financial crisis. Irrigation subsidies at that time totaled nearly US$500 million, and the budget squeeze in subsequent years drastically curtailed funds for investments in, and maintenance of, irrigation networks (OECD 1997). By the end of the 1980s, the deterioration of Mexico's irrigation infrastructure had begun seriously to affect the productivity of Mexico's irrigated land (World Bank 1995; Kloezen 1998). The government concluded that the best—perhaps the only—way to sustain irrigated agricultural production was to transfer control of water to local groups of farmers and to make them responsible for raising most of the revenue for maintaining the irrigation systems. The underlying assumptions were that local groups would best understand local conditions, and that they would be willing to pay most of the maintenance costs if the resulting benefits were clearly visible.

By 1998, irrigation management for about 95 percent of Mexico's publicly irrigated land had been transferred. (Management for more than 2.5 million hectares of privately irrigated land was largely unaffected.) Local user costs covered about 90 percent of the upkeep of these systems, which also meant that the irrigation systems were now largely insulated from the vagaries of the federal budget (Johnson 1997). In 1995, for example, irrigation modules were little affected by the macroeconomic crisis that had a devastating effect on many line items in the federal budget.

The transfer of control also had the effect of streamlining operations. Farmer groups proved themselves much more demanding of irrigation personnel, and during the first five years of the transfer program, local irrigation staffs were cut an average of about 50 percent (Johnson 1997). Federal subsidies on maintenance dropped to zero by 1993, yet maintenance of the systems had greatly improved.

As one of the early test cases, the Yaqui Valley was very much a leader in the decentralization of water management. After receiving the concessions, Yaqui units raised user charges to farmers. For the period 1988–1994, the real cost of water to farmers rose by about 70 percent. Fortunately for farmers of the Yaqui Valley, the water flows of the Yaqui irrigation system were fairly uniform through the mid-1990s at about 2 billion cubic meters of water annually. Farmer concerns about being given concessions for only the percentage of the water flow deemed appropriate by the CNA (National Water Commision), therefore, did not create reservoir management difficulties in the first half of the decade.

As discussed in chapter 11, however, the drought of 1996–2004 caused serious problems for Yaqui Valley farmers. Water prices rose steadily over this period, and in the 2003–04 crop cycle—after the Yaqui reservoir had been drained dry—the price of irrigation water suddenly doubled.[11] The costs of irrigation water and drainage rose to 15 percent of farmers' variable costs, just below the cost shares for fertilizer and pest control, and remained as high through 2005–06.

The play between biophysical constraints on production caused by drought on one hand, and policy incentives supporting production of crops that were not drought tolerant on the other, has been interesting to observe since 2000. With water now back in the reservoir, water prices to farmers have fallen and the dominance of commodity price policy over water resource policy appears to have gained an upper hand once again.

Toward Freer Trade

The transition from coupled price supports to decoupled income supports, along with the liberalization of factor markets, enabled Mexico to move toward a regime of freer trade for the agricultural sector. NAFTA, signed in January 1994, was the first trade agreement between advanced industrial countries and a developing country in which most of agricultural trade was included (OECD 1997). A key motivation for entering NAFTA was to increase levels of direct foreign investment and trade flows; in particular, Mexico expected to gain from foreign investments in the manufacturing and export industries (Yúnez-Naude and Taylor 2006). Although NAFTA encompassed trade in all sectors of the Mexican economy, it had particular relevance for agriculture. In 1993–94, almost 90 percent of the country's farm exports were shipped to the United States and Canada, and over three-quarters of its agricultural imports came from these two countries.

The trade liberalization measures associated broadly with NAFTA had important consequences for both the level and stability of farm incomes in Mexico. Prior to the 1990s, effective import monopolies for many commodities shielded the agricultural sector from substantial overvaluation of the peso and from the fluctuations of international market prices. With freer trade, agricultural producers—particularly in commercial regions like the Yaqui Valley—became more vulnerable to macroeconomic policy swings and global market volatility. At the same time, Yaqui farmers benefited from the devaluation of the peso during 1994–96 (discussed later) and the consequent gain in international competitiveness for their crops. In short, farmers no longer operated in a protected, semiclosed economy, and their private profits at the turn of the century were subject to both the benefits and hazards of international price movements.

The links among trade, macroeconomic, and agricultural price policies during the 1990s underscore how the interaction of multiple reforms affected farmers in the Yaqui Valley. While farmers often blamed NAFTA for changes in efficiency and equity within the agricultural sector, it represented only one of many reforms in the Mexican economy. Moreover, a series of policy and economic shocks also altered the profitability of farm production. Despite these shocks, the directions of change in the agricultural sector—and in the rest of the Mexican economy—would likely have been very different during the 1990s without Mexico's vigorous move toward freer trade.

Unexpected Policy and Economic Shocks

The foregoing review of agricultural reforms in the Mexican economy indicates that policies designed for a nation often affect particular regions, such as the Yaqui Valley, in quite specific ways. They are thus largely exogenous to the agricultural region in question. Other exogenous economic forces—most notably national macroeconomic policies that alter exchange rates and interest rates—can also have large and differential effects at the regional level. This phenomenon was never clearer to Yaqui Valley farmers than during the peso crisis in the fourth quarter of 1994. In addition, Yaqui Valley farmers were battered by exogenous swings in international commodity prices.

Macroeconomic Shocks

The peso crisis in December 1994 originated seven years earlier when the Salinas government launched a stabilization program to reduce double-digit inflation. The program centered on strict exchange rate controls,[12] tight fiscal and monetary policies, and liberalization of financial markets to attract foreign investment. By the end of 1993, just before NAFTA went into effect, the annual inflow of foreign capital exceeded US$29 billion, and consumer inflation fell to an annual rate of 8 percent. At the same time, however, real annual interest rates on peso-denominated short-term bonds (*cetes*) were in the 12–16 percent range; real GDP growth remained low (only 2.7% on average between 1988 and 1993); and the current account deficit rose to over 6 percent of GDP (Ramírez 1996; Savastano et al. 1995). Those trends, which occurred within the context of strict exchange-rate controls, led to an effective appreciation of the peso by more than 60 percent between 1987 and 1992 and to a further appreciation of 32 percent in 1993 (Savastano et al. 1995; Puente-González 1999).

The causes for the specific timing of the peso collapse remain a contentious issue. There can be little doubt, however, that in 1994 the exchange rate was overvalued, the economy was stagnating, and the growing current account deficit was not sustainable. High interest rates and an artificially strong peso were creating difficulties for domestic industry. Moreover, a series of destabilizing political events in 1994 created substantial investor uncertainty (Ramírez 1996). Portfolio investors, who had invested almost nothing as late as the first quarter of 1991, increased their quarterly financial flows to almost US$8 billion by the fourth quarter of 1994. In December 1994, however, they withdrew massive sums in the course of only a few days.[13]

A major panic at this time was set off by the administration's move to widen the exchange-rate band and then to allow the exchange rate to float. The failure of the Mexican authorities to consult adequately with the international financial community prior to the float almost surely aggravated the peso panic.[14]

The result was the collapse of the peso. The M$/US$ rate, which stood at M$3.4/US$1 on December 1, 1994, fell to M$6.0/US$1 on January 31, 1995. The crisis was both good news and bad news for the farmers in the Yaqui Valley. They were affected negatively by the associated high real rates of interest and higher prices on imported farm inputs, but they benefited greatly from the increased competitiveness of their export crops.

In the decade following 1995, the ministry of finance relied to a greater extent on monetary aggregates and a floating exchange rate to restrict demand and control inflation (Nadal 2003). As shown in table 8.1, this approach led to higher interest rates, which relieved some pressure on the exchange rate but also reduced the flow of credit to productive activities in Mexico. Interest rates fell into the single digits between 2002 and 2006, and commercial bank credit began to expand once again (*Economist* 2006). Overall, Mexico maintained a semifixed and slightly overvalued exchange in order to curb inflation. Exchange rate stability, in turn, boosted the confidence of foreign capital holders (Nadal 2003), although the currency value had a somewhat dampening effect on exports. One of Mexico's continuing challenges is to remain competitive internationally, particularly with respect to exports to the United States. This challenge has become ever more pressing since China's entry into the WTO.

World Price Shocks

The ramifications of the peso crisis in 1994–95 were widespread in the Yaqui Valley, but perhaps no more so than the sharp rise and the even sharper decline in world commodity prices that characterized the 1990s. Mexico's new agricultural openness meant that Yaqui Valley farmers, other things equal, benefited when world prices rose for their commodities. A

TABLE 8.1 Mexico, macro prices, 1990–2006.

	Short-term interest rate	*Nominal exchange rate (peso/US$)*	*Annual inflation (CPI)*
1990	35.0	2.8	26.7
1991	19.8	3.0	22.7
1992	15.9	3.1	15.5
1993	15.5	3.1	9.8
1994	14.6	3.4	7.0
1995	48.2	6.4	35.0
1996	32.9	7.6	34.4
1997	21.3	7.9	20.6
1998	26.2	9.2	15.9
1999	22.4	9.6	16.6
2000	16.2	9.5	9.5
2001	12.2	9.3	6.4
2002	7.5	9.7	5.0
2003	6.5	10.8	4.5
2004	7.1	11.3	4.7
2005	9.3	10.9	4.0
2006	7.3	10.9	3.6

Source: OECD 2007

significant improvement in world prices was precisely what happened in the period from 1991 to 1996—the increase beginning fortuitously just at the time when the reforms were being implemented. World prices (in nominal US$) for wheat, maize, and soybeans prices *rose* between 1991 and 1996 by 60, 53, and 33 percent, respectively (fig. 8.1). Farmers reported difficulties in sorting out reform effects from global price movements, but rising world prices for agricultural commodities helped greatly to offset the reductions in crop-specific subsidies within Mexico.

As farmers throughout the world know, however, prices that go up can also go down. Between 1996 and 1999, world prices *fell* by 44, 41, and 32 percent for wheat, maize, and soybeans, respectively (fig. 8.1). Differential inflation rates in Mexico relative to world inflation and the continuing depreciation of the Mexican peso relative to the US dollar complicate comparative peso profit calculations through time (Naylor et al. 2001). There can be no doubt, however, that globalization added to price instability for Yaqui Valley farmers, and that in the last half of the 1990s, falling

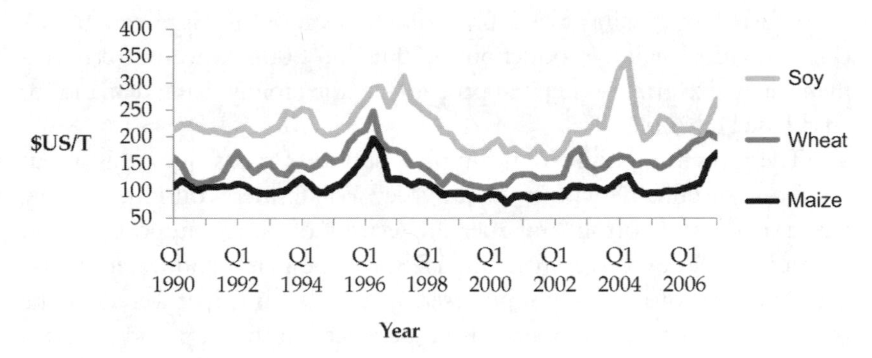

FIGURE 8.1 Nominal world prices (IMF 2007).

commodity prices were significant factors in putting farmers into a difficult cost-price squeeze.

Renewed Commodity Protection

Economic and policy changes affecting agricultural decision making in the Yaqui Valley in the 1990s were largely exogenous to the farming system. Farmers in this highly commercialized region of Mexico are not just "policy takers," however. They also *do* have a voice in the development of national agricultural policies. Their generally weak competitive position relative to the United States, coupled with the fall in international commodity prices in the latter part of the decade, induced an intensive lobbying effort that ended with the reenactment of commodity price supports in 2000 for bulk crops such as wheat, maize, and cotton. Yaqui farm groups were not solely responsible for the renewed protection, but they had—and continue to have—a persuasive influence on other farm groups in Mexico and a history of political power in Mexico City. Three Mexican presidents came from Ciudad Obregón, and the connection between large farm organizations in the Yaqui Valley and politicians in Hermasillo (the state capital) and Mexico City have traditionally been strong. When lobbying has not worked, farm groups have resorted to other tactics, such as threatening to close down the highway that runs through the state of Sonora to the US border.

The policy shifts toward protection at the turn of the century have been quantitatively significant. From a low of 5 percent in 1996, the Mexican agricultural PSE reached a modern-day high of 26 percent in 2003 (OECD 2007). By 2006, however, the PSE had fallen again, mainly as a consequence of a rising world price of wheat caused in large part by declines in global wheat production in 2005 and 2006 and upward movements in maize prices generated primarily by the rapidly rising demand for corn-based ethanol.

The growth of explicit and implicit subsidies raises questions about the future sustainability of the Yaqui Valley. Will farmers continue to grow wheat, corn, and cotton to garner the security of policy support, despite the fact that these crops demand high fertilizer, pesticide, and water inputs? If so, will improvements in input use efficiency be sufficient to reduce the cost-price squeeze that farmers have experienced in the past, and to lessen the impact on the environment? Will farmers continue to band together in the promotion of these crops in order to preserve their "safety in numbers"? Answers to these questions will depend on agricultural policy dynamics within Mexico—driven in large part by the tension between northern commercial interests versus southern social interests—and between Mexico, the United States, and Europe. As long as the United States and the European Union persist in subsidizing wheat, maize, and cotton through various policy instruments, Mexico will be forced politically to subsidize these crops as well. It is unlikely that Mexico will even consider dismantling commodity price supports in Mexico in the near future (Luis Signoret, president of the Association of Producer Organizations in Southern Sonora,[15] interview, 2006). The future course of commodity protection in Mexico depends importantly on progress on agricultural trade issues within the WTO, on US farm policy, and trends in international commodity prices. Of particular importance is how the price of crops, especially cereals and oilseeds, will be affected as a result of the development of biofuels. Although Yaqui Valley farmers are unlikely to supply maize or wheat directly to biofuel plants, farmers in the region could stand to gain from cereal prices that, depending on the price of oil, are determined by new links to the energy sector (Naylor et al. 2007).

Farmers' Responses to Changed Circumstances

An important part of our Yaqui Valley story lies in the response of farmers to these multiple policy changes, biophysical shocks, and international price swings. Have farmers been able to adjust to the changed circumstances in ways that portend long-run stability for agricultural incomes? What have been the equity and environmental consequences of these changes? In order to address these questions, researchers at Stanford University and

CIMMYT conducted a series of farm surveys between the early 1990s and mid-2000s, that focused on changes in farm practices, agricultural costs and returns, and resource use in the Yaqui Valley.[16] The surveys were based on stratified random samples of producers in the valley and included between 50 and 120 interviews each. The interviews were conducted in Spanish at the farm sites and within *ejido* villages. Data from the surveys were then analyzed using various statistical techniques to determine patterns of correlation and causality, and to measure changes in financial and economic profitability over time. In addition, Landsat satellite images were analyzed to examine land-use changes in the Yaqui Valley (Luers 2004; Luers et al. 2006; Lobell et al. 2004).

Overall, the surveys showed that producers reacted to fluctuating price, policy, pathogen, and weather factors by altering production patterns and farm management practices, exploring new marketing and credit arrangements, and expanding the size of operations. But as chapters 9 and 11 also show, farmers' responses to changes in fertilizer prices were limited, and their planning methods for dealing with decreased water availability in the 1992–2004 period were inadequate.

Changes in Crop and Animal Production

Changed cropping patterns were farmers' most obvious response to the changing circumstances, particularly in the areas planted to wheat and soybeans. As indicated in figure 8.2, virtually all of the wheat grown in the Yaqui Valley during 1991 was bread wheat. By contrast, more than 80 percent of the wheat grown in 1996 was durum wheat. This switch was driven partly by price and partly by karnal bunt, which affects bread wheat predominantly (Wilcoxson and Saari 1996). Similarly, soybeans went from being the dominant summer crop in 1991 to occupying nearly zero area in 1996 because of the whitefly invasion (fig. 8.3). From a farm planning perspective, the loss of soybeans was one of the most serious problems of the 1990s. A number of key crops such as cotton and autumn-planted maize do not "pair" annually with a winter-planted bread or durum wheat crop because of seasonal overlaps, and thus annual farm incomes fell.

Another striking story told by figure 8.2 is the rise in durum wheat production relative to the other major crops. Although wheat is not a particularly drought-tolerant crop, it continued to be the main crop in the valley. Wheat yields remained high, and farmers adjusted their practices to become slightly more efficient in fertilizer use and water use, as discussed

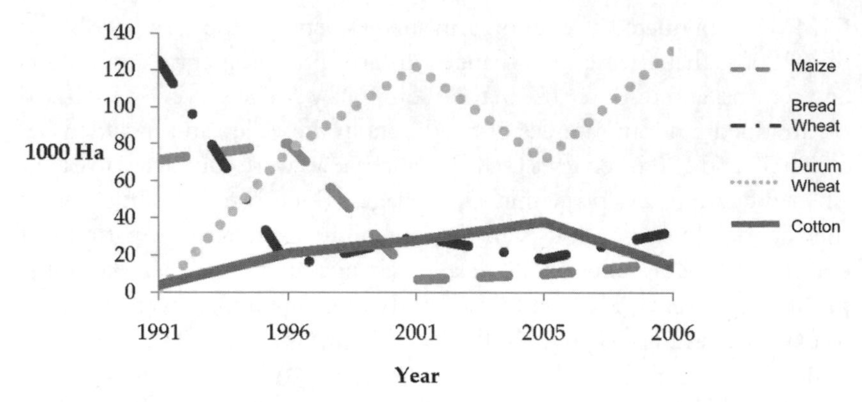

FIGURE 8.2 Harvested area, Yaqui Valley fall crops (SAGAR, various years).

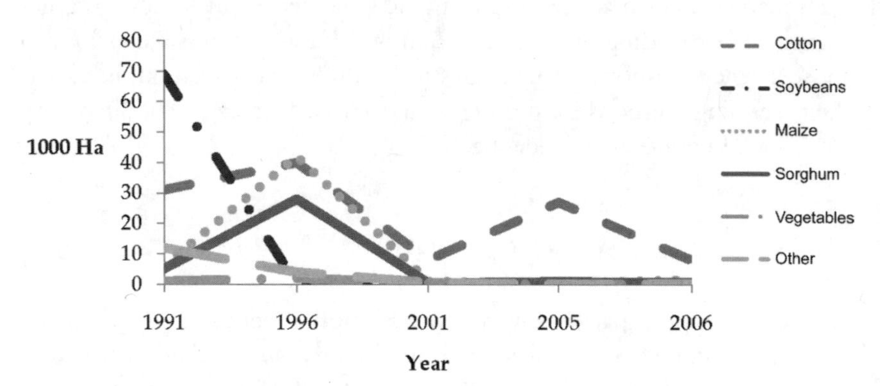

FIGURE 8.3 Harvested area, Yaqui Valley spring/summer crops (SAGAR, various years).

in subsequent chapters of this volume. Success in keeping costs for wheat relatively low in comparison to maize and cotton was a key factor in determining the crop rotation.

Three other changes in output patterns are also worthy of note. First, a combination of Mexico's new economic openness, the increased urbanization of the Yaqui Valley, and rising per capita incomes had a positive, although marginal, effect on the production of higher-valued crops. The Yaqui Valley had long been known as a producer of bulk commodities. On several earlier occasions, farmers produced melons, only to suffer disastrous consequences when increased supplies caused domestic melon prices to plummet and induced US authorities to impose import restrictions at the border. During interviews, many farmers continued to express skepticism about the potential of demand-driven products—a skepticism that was reflected in the very limited area in the early 1990s devoted to such products. The aggregate area in vegetables was still not large in 1995–96, but by 1998, area under vegetables had doubled relative to 1991 (SAGAR 1998). Citrus groves also doubled during the 1990s.

A second economic and agricultural force in the valley was growth of the livestock sector. Livestock numbers grew quite rapidly between 1991 and 1996, and by the latter year, production of pigs and poultry was about 350,000 head and 6 million birds, respectively (SAGAR 1998). The competitiveness of the livestock sector clearly improved as a result of the changed subsidy policies on grains associated with a decoupling of price supports in the early 1990s. New price relationships, in turn, induced a restructuring of the Yaqui Valley pork industry into much larger units (Southard 1999). This shift into livestock also had consequences for regional feed grain markets. Wheat, for example, became increasingly important as a local livestock feed, taking up to half of the Yaqui Valley's wheat output by the end of the 1990s. The feed use of wheat, both in Mexico and globally, may increase further if maize-based bioethanol were to expand, thereby pushing maize prices upward to a new plateau (Naylor et al. 2007).

A third effect of the reforms was the emergence of aquaculture—particularly shrimp farming—as a significant economic activity (Luers et al. 2006; Villa-Ibarra 1998; chap. 7). The southern Sonora coastal zone, of which the Yaqui Valley is a part, was a relative latecomer to shrimp farming, but between 1994 and 2001, 10,000 hectares of coastal land were converted to shrimp ponds (Luers et al. 2006). This increase was ten times as great as the total land area converted between 1988 and 1993. Data from the 1998–99 Stanford University survey indicated that coastal land quadrupled in value during the second half of the 1990s.

Research by Luers (2004, 2006) also showed that as a result of the land-tenure law reforms, many *ejido* communities gained legal title to their land, divided communal land into private parcels, and rented or sold their land to private landholders. This process varied spatially along the coast depending on when the land was first reformed to *ejido* title. *Ejiditarios* remained active in shrimp farming, although there was a marked shift from ownership to rental; in 1996, 70 percent of coastal *ejidos* owned their land, and by 2001 the share had dropped to 30 percent. One of the main reasons for this shift was that *ejidos* had limited access to credit and technical support for their shrimp operations. Exceptions to this trend were the *ejido* communities that gained access to land under the Salinas administration and received initial government support for aquaculture development.[17]

Overall, changes in, and consequences of, land use and tenure along the coast were not uniformly distributed in Sonora but instead varied with the historical, geographical, and socioeconomic characteristics of the specific *ejido* communities. Our study demonstrated that despite an almost unanimous interest in entering shrimp farming, less than 10 percent of the *ejidos* in southern Sonora participated in the post-reform shrimp farm development. If the patterns of shrimp farm growth and shifts in land tenure characteristic of the late 1990s and early 2000s continue, many communities in the *ejido* sector will likely find that their hopes of earning riches from shrimp farming have disappeared along with much of their traditional coastal lifestyles.

Altered Credit and Marketing Arrangements

Clearly, the opening up of the Yaqui Valley by the reforms affected product markets, factor markets, and the welfare of various groups in ways that were not easily predicted in 1991. Credit availability played a significant role in producers' ability to remain in business; it also influenced production practices in the valley. In particular, institutions governing credit for, and sales of, exportable products changed in response to macroeconomic conditions in the mid- to late 1990s. During the 1994–95 cropping season, farmers were faced with interest rates that soared to 80 percent in nominal terms (30% in real terms) and an exchange rate that had lost half of its value. While the latter was beneficial for exports, interest payments became one of the largest variable costs in farmers' budgets (Naylor et al. 2001). Many small-scale farmers simply could not pay the interest charges and lost their ability to obtain loans for future seasons (Lewis 2002).

By the 1995–96 winter cropping season, however, new lines of credit became available, which were arranged through producer organizations in the Yaqui Valley and enabled farmers to obtain credit in dollars in exchange for contracts to sell most of their output in dollars (with specified quality standards). Farmers who secured credit in dollars were thus able to lower costs and eliminate much of the exchange rate and interest rate risks associated with fluctuating macroeconomic conditions. Unfortunately, *ejiditarios* and other small-scale farmers were largely excluded from dollar-based contracts. Some subsidized credit remained available for these groups, but it was reduced significantly, and many small-scale farmers were unable to secure credit at either subsidized or unsubsidized rates (Lewis 2002).

The emergence of new credit and marketing institutions thus had important implications in determining who was doing the actual farming in the Yaqui Valley, and what they were doing in terms of farm management. Moreover, our most recent surveys suggested that fertilizer applications, water use, and land use all hinged to some considerable extent on the advice of credit institutions (chaps. 3 and 5). There is evidence, for example, that farmers did not feel at liberty to deviate from credit-based recommendations on fertilizer use, and as a consequence used more of that input than was necessary (chaps. 3 and 5).

Changed Farming Practices

Policy reforms and macroeconomic shocks during the 1990s had major impacts on farm practices. The cost structure of farming in the Yaqui Valley changed dramatically between 1991 and 1996 (Naylor et al. 2001). Fertilizers became more costly with the elimination of subsidies, machinery costs increased with the decline in energy subsidies and the rise in import costs, and expenditures on interest escalated with changes in macropolicy. As a result, many farmers adjusted the quantity, quality, or method of application of their inputs. Our surveys revealed several interesting outcomes. First, although high rates of nitrogen use were attributed in the prereform period to subsidized prices, farmers did not respond to price increases after the reforms by lowering their fertilizer rates. The number of fertilizations increased on average over the decade from 1991 to 2001 from 2.3 to 2.8 per season, and the total amount of nitrogen per hectare applied per season did not drop.

Instead, rising costs induced farmers to adjust other practices besides fertilizer applications. Land preparation in the Yaqui Valley, which typically involved a series of machinery passes (for example, six to seven for wheat), became more efficient. A large number of farmers eliminated at least one pass per cycle during the 1990s in response to rising fuel prices and farm machinery costs (Naylor et al. 2001). They also reduced the number of postplanting irrigations from four to three. However, the amount of water used per irrigation remained largely unchanged, and the preplanting irrigation was maintained to control weeds. Farmers also moved to the use of beds for planting their wheat crops. The use of such beds improved irrigation water efficiency, facilitated mechanical weeding, and permitted more precise placement of fertilizer. In the future, farmers' ability—and willingness—to improve input use efficiency further will depend on their perceptions of the outcome, as well as the advice given by technical experts, many of whom appear to be linked to credit organizations (chap. 5).

Expanded Operational Holdings

A final important outcome of policy reform and macroeconomic change in the 1990s was an increase in operational holdings in the valley, which resulted mainly from an increase in the amount of *ejido* land rented out to private landholders. In the survey on land rental markets in 1998–99, 70 percent of the *ejiditarios* interviewed were renting out their land, and 96 percent of these rentals were to the private sector (Lewis 2002). Almost all of the *ejiditarios* renting out their land reported that limited access to credit was a primary motivation. Other reasons for renting out their land included high input prices, high water prices, insufficient water supplies, low crop prices (especially with the elimination of crop price supports on key crops), and a shortage of affordable machinery. The survey results indicated further that the growing trend in rentals by *ejiditarios* had already eroded the cohesiveness of *ejido* communities in the Yaqui Valley (see table 6.1 in chap. 6; Lewis 2002).

The *ejido* communities were not the only ones affected by the policy changes and shocks of the 1990s. Some private landowners also sold or rented out their land. The expanding sales and rental market led to an increase in operational holdings for larger farmers in the Yaqui Valley and a consolidation of farming activities. Surveys showed that in 1991–92, about 55 percent of the wheat area was farmed by *ejidos*,[18] 37 percent of the wheat area was on land owned and farmed by the private sector, and 8 percent was produced on land rented by the private sector. A decade later, in 2001–02, the respective numbers were 33 percent, 21 percent, and 46 percent. This trajectory has not yet stabilized. It appears, however, that policy reforms, especially in Article 27, increased the scale of operational units in the valley and hastened the demise of numerous *ejido* communities as agricultural producers.

The general trend during the 1990s and early 2000s was that the big got bigger and the small suffered, although the Yaqui Valley still pales in comparison to the United States in terms of economies of scale and operational holdings. Indeed, further consolidation may be the only way that Yaqui farmers can maintain competitiveness with bulk commodities vis-à-vis farmers in the United States, Canada, and Europe in the long run. The interactions between land, credit, and labor markets will also shape the long-run welfare outcomes in the valley. With increased migration of farm household members to the United States and a movement of labor out of agriculture more generally, large farmers increasingly have complained of difficulties in finding labor to work their fields. If the pattern of consolidation continues in the valley, a possible result will be a migration of farm workers from southern states in Mexico and southern countries in Latin America to Sonora—at the same time that Sonorans themselves are migrating further north.

Changing Profitability of Farming in the Yaqui Valley

The combined forces of policy, price, biophysical, and behavioral changes affect private (financial) profits—the market costs and returns to crop production—and thus the welfare of farmers in the Yaqui Valley. They also influence social (economic) profits—the net returns to society as a whole from crop production. By assessing the latter, it is possible to derive broad conclusions about the efficiency impacts of public policy at both micro- and macroscales (Monke and Pearson 1989).

Figures 8.4a and 8.4b show real (inflation-adjusted) per-hectare costs and returns for the major crops in the Yaqui Valley in 1991, 1996, 2001, and 2006.[19] The results are derived from "synthetic" estimates based on the surveys and a substantial number of federal, state, and district reports (Puente-González 1999). Profits are reported in financial prices (what farmers actually paid and received) and in economic prices. The latter prices value tradable products and tradable factors at world prices, and nontradable factors, such as labor, in terms of their domestic opportunity costs. These estimates also include the rental cost of land in total costs and the marginal cost of irrigation water, that is, the cost per hectare of operating and maintaining existing irrigation systems. Lastly, the economic calculations use a shadow exchange rate to correct for any overvaluation of the peso.[20]

Farm profitability in 1991 was heavily dependent on the degree of protection given the grain sector by government policy. As shown in figure 8.4a, wheat, fall-planted maize, and cotton were financially very profitable; however, from an efficiency point of view this profitability was largely illusory. If farmers had been forced to compete at international prices, they would have sustained very large losses. Not all grains were given this protection, and the financial and economic losses indicated for sorghum show why that crop was seldom grown in the valley. Of the shorter summer

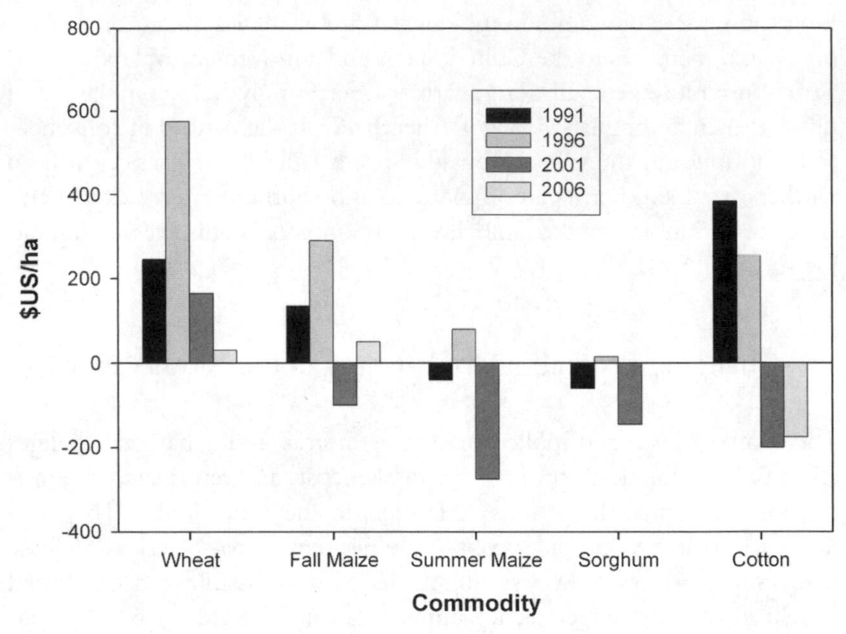

FIGURE 8.4a Real financial profits ($US/ha), 1991–2006 (Stanford surveys; SAGAR [various years]; Naylor et al. 2001).

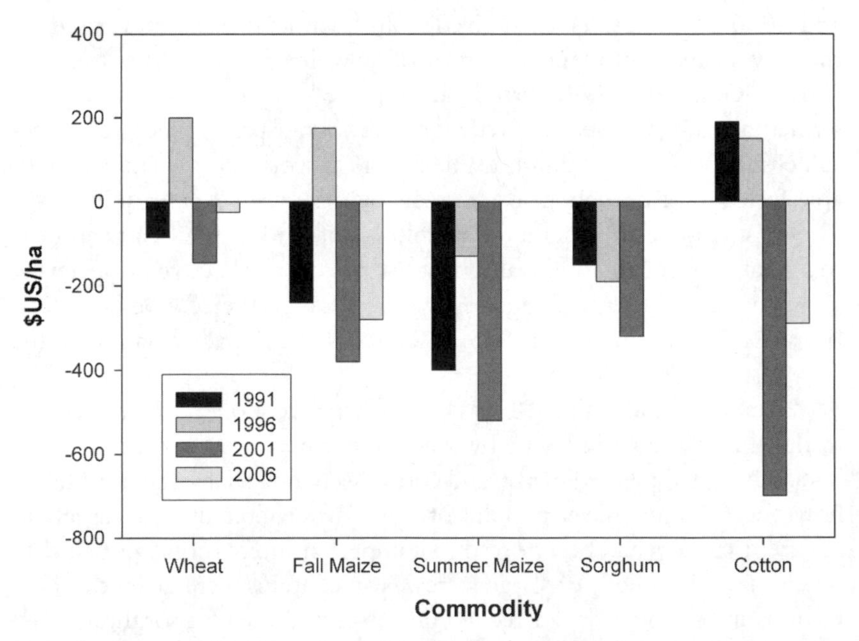

FIGURE 8.4b Real economic profits ($US/ha), 1991–2006 (Stanford surveys; SAGAR [various years]; Naylor et al. 2001).

crops, only soybeans (not shown in the figure)[21] were financially profitable. With grain prices generally guaranteed, most farmers opted for the wheat/soybean combination, although cotton was also a viable alternative. The problem with cotton is its awkward timing in the rotation; it is typically grown from early March through September and thus cannot be grown in an annual cycle with wheat or soybeans.

Just five years later, the rural economic situation had changed dramatically. Yields for most crops were up substantially in 1996 as a consequence of better seeds, improved water management, and additional inputs, as farmers responded to higher prices. At that time, durum wheat had largely replaced bread wheat, and soybeans had dropped from the rotation. Even with high world prices, farmers in the valley had few profitable alternatives among the summer crops. With few opportunities for double cropping, the basic choice faced by farmers was a fall-planted grain crop or a spring-planted cotton crop (fig. 8.4a). The greatest profit potential was in the autumn season, with wheat dominating the profit margin. Moreover, surveys indicated that farmers became increasingly aggressive in trading summer claims on irrigation water for assurances of enough water to sow 100 percent of their land during the fall season.

Perhaps most important was the change toward positive economic profitability for grains in 1996 relative to 1991. The reduced distortion occurred in large part because of the reforms. Direct subsidies to the agricultural sector were greatly reduced in 1996 with the replacement of coupled price supports by decoupled income supports, and implicit subsidies were reduced by the devaluation of the peso. Even more significant, however, was the sharp increase in international grain prices. As shown in figure 8.1, the upward spike in world grain prices probably gave farmers a false sense of security, as the post-1996 era was soon to demonstrate.

A major turning point occurred in the Yaqui Valley in 2001. Farmers faced a prolonged drought, relatively low international commodity prices, and high costs. US farmers were outcompeting Yaqui farmers in most crops—most notably wheat and maize—and were receiving a variety of indirect subsidies for grain production (e.g., emergency relief payments) that Mexican farmers did not receive. As a result of these pressures, and pressures by Mexican growers in response to the competition, the government reinstated crop subsidies on all of the major crops, as discussed earlier. Despite these subsidies, wheat was the only crop that remained financially profitable in the valley in 2001 (fig. 8.4a). The sad fact was that no crops were economically profitable at this time (fig. 8.4b).

In the most recent period, grain prices rose once again in world markets, allowing both wheat and fall-planted maize to become financially profitable. Summer maize dropped out of the rotation, and sorghum was not widely planted. As in 2001, no crops were economically profitable. This pattern is particularly worrisome in terms of the economic sustainability of crop production in the Yaqui Valley. If crop production can remain profitable only with government subsidies and other protective policies, is the system really sustainable? Is it possible for developing countries like Mexico to avoid subsidizing agriculture as long as the United States and Europe continue to subsidize agriculture? Moreover, since wheat and maize use large amounts of irrigation water relative to some other crops like chickpea or drip-irrigated vegetables, is the system even biophysically sustainable? Unless Yaqui farmers can move toward niche markets, or unless world commodity prices rise considerably (as they have in 2010–11, and remain so), agricultural sustainability in the valley remains in question.

Environmental Costs and Benefits

The economic profitability calculations shown in the previous section are based on adjustments of financial costs and returns for policy distortions. A full accounting of social profits should also include adjustments for environmental externalities—the true resource costs of production. Unfortunately, such accounting is extremely difficult to do in most situations, including the Yaqui Valley. In order to determine monetary values of environmental externalities, three steps are required. First, the environmental effects of production must be quantified in biophysical terms. As this volume demonstrates, an enormous amount of research is needed to determine effects such as nitrogen loss to the environment from different production practices, changes in hydrological cycles from continuous irrigation under variable climate conditions, and other environmental effects. The second step is to measure the impacts of these environmental effects on ecosystem function, human health, and other indicators of "welfare" in the system. Much broader and longer-run ecology and epidemiology studies than have been done in this study are needed for better than crude attributions. Finally, monetary values need to be placed on welfare changes. These values, which include, for example, the value of human lives and species preservation, can be done with various empirical techniques (Kolstad 2000) but are nevertheless subjective. As a result of these difficulties, we limit discussion to qualitative assessments of the environmental costs and returns to the system, and discuss policy measures designed to address them.

The balance between development and environment has been skewed toward the former throughout the Yaqui Valley's history of intensive agriculture. High rates of fertilizer use and losses to the atmosphere and groundwater (Rice 1995; chaps. 3 amd 10); pesticide pollution; massive hydrological diversions affecting both terrestrial and marine ecosystems; wasteful irrigation practices stemming from inadequate pricing; extensive crop residue burning; and intensive tillage practices all add to a somewhat bleak picture of environmental and human health in the valley. For our study, Kitzes (2005) reviewed the limited literature on human health impacts of wheat agriculture in the Yaqui Valley. Similar assessments are still needed for cotton, maize, livestock, shrimp, and other major production activities.

The results of Kitzes's work show that wheat production imposes measurable costs associated with climate change (specifically, emissions of nitrous oxide [N_2O] from fertilizer use and methane [CH_4] from livestock, both of which are greenhouse gases), human and ecosystem pesticide poisoning, respiratory disease from emissions from crop residue burning and soil fertilization, and premature mortality. The costs of hospital visits due to respiratory problems and pediatric care for asthma related to particulate matter from crop residue burning are tabulated in Kitzes's study, and premature mortality stemming from respiratory problems is estimated using the US Environmental Protection Agency's (EPA) framework. Although these methods contain considerable margins of error, the overall message is clear: the value of respiratory disease and premature mortality associated with agricultural practices, especially the burning of straw, is substantial. Using blunt (and somewhat controversial) measurement instruments, the cost is comparable to the sum total of private profits in the Yaqui Valley (Kitzes 2005). What is striking about this work is that the environmental costs of wheat production from residue burning alone would make the crop even more *unprofitable* in social terms than it is currently. The Yaqui Valley has initiated burning regulations, involving fines of about US$75/ha, intended to reduce smoke and particulate-matter pollution. In addition to increased baling of straw, considerable inroads have been made on residue management via low-tillage farming practices (Harris 1996); however, the costs of dealing with the straw associated with six-ton wheat yields are major problems both for wheat farmers and for the Yaqui society as a whole.

Mexico's Ministry of the Environment and Natural Resources (SEMARNAT) has introduced several policies aimed at reducing environmental threats from intensive crop and animal production, although the enforcement of these policies remains elusive. The key laws of interest to the Yaqui Valley include regulations of pollution discharge to water bodies and regulations of use of pesticides and other toxic substances.[22] In addition, there are health standards for laborers working with pesticides, similar to the US Occupational Safety and Health Administration (OSHA) laws.[23] Of particular concern to commercial agricultural regions like the Yaqui Valley, however, is the documented substitution of persistent, less acutely toxic pesticides with less persistent, but more acutely toxic, pesticides on export crops (Albert 1996). Due to importing countries' concern over food safety, Mexican field workers have suffered higher health consequences, despite efforts by some producers to counteract such substitution. Finally, there are federal and local regulations prohibiting the burning of crop residues, particularly in areas where forest fires related to crop burning are likely, but burning remains a common practice throughout much of the country.

A large part of the problem in Mexico—as in many other countries around the world—is that the budget allocation for environmental protection is substantially lower than for economic development and export promotion. This problem is particularly serious for Mexico relative to most industrial and other Latin American countries (OECD 2003). SEMARNAT has traditionally been accorded less decision-making power, smaller budgets, and fewer resources than other Mexican ministries; for example, in 1999 it received less than 6 percent of overall federal budget allocations (Varady et al. 2007). Over three-quarters of this budget allocation was for the CNA, leaving 1.3 percent of the total federal budget for environmental protection under SEMARNAT (Romero-Lankao 2000).

An increasing share of the oversight for environmental protection has now been transferred to the state level. Given the challenges that producers face in Sonora in trying just to remain competitive and survive financially, it is unlikely that strong environmental regulations will be implemented and enforced anytime soon. For example, according to existing laws, pork producers in Sonora were expected to be fully compliant with environmental regulations on water discharge, solids disposal, and odors by 2008. These producers are struggling to maintain competitiveness with the United States after the 15 percent tariff on pork from the United States was removed as part of the NAFTA agreement. Moreover, they face the additional disadvantage that many US pork producers are paid to clean up their wastes through the Environmental Quality Incentives Program (EQIP) of the US Farm Bill. The compliance period in Sonora will likely be delayed, but only at the expense of environmental quality.[24] The financial disadvantage that producers of all commodities in the Yaqui Valley experience relative to their counterparts in the United States and Europe continues to threaten the outlook for both economic and environmental sustainability in the valley.

Conclusions, Opportunities Missed, and Lessons

This chapter raises serious concerns about the long-run economic viability of production of bulk crops in the Yaqui Valley. If direct price supports are withdrawn in Mexico, and if international commodity prices do not rise substantially above the levels recorded in the late 1990s and early 2000s, it appears that Yaqui farmers will face three main options. In this respect, they are typical of most farmers throughout the world (Dixon et al. 2001). The first option is to improve input use efficiency to a significant degree in order to lessen the cost-price squeeze that farmers have experienced in the post-1996 era. Farmers' ability and interest in such improvements is discussed in detail in chap. 3. A second and more difficult option is to diversify into higher value, niche-market crops. Some farmers have moved in this direction during the past decade, but they have experienced swings in climate—including more frequent freezes and reduced rainfall—which have driven them back into the bulk crops. Many farmers worry that there is no research support on the technical side for such diversification. Moreover, the market can handle only so many melons or green beans before prices fall. After decades of growing staple commodities like wheat and maize, it may be that farmers are constrained by the idea of diversification itself.[25] A third option, of course, is to leave agriculture altogether.

In presenting these potential options, it has become clear that our study was limited in several dimensions in the policy area. While diversification may be the key to Yaqui farmers' future success and long-run sustainability, our study did not fully explore the opportunities and limitations of this path. We did not have the appropriate colleagues as part of our team—business experts who could analyze the market potential of diversification and create business strategies. We also did not allocate the time or resources to study the biophysical constraints on high-valued crop production, particularly climate, pest, and disease risks. Most of our study focused on wheat production, just as most producers in the valley continue to focus their attention on wheat production. We may have thus missed an opportunity to explore whether the valley was in a position to take advantage of forward contracting to large supermarkets as part of the "super-market revolution" that is of growing importance in middle-income countries (Reardon and Berdegue 2002).

In addition, our study did not fully explore the social repercussions of constitutional reform, macroeconomic and biophysical shocks, and land consolidation. While we understood broadly that *ejido* communities rented and sold an increasing share of their land, and that they struggled to remain in business due to a lack of credit and institutional support, we did not perform studies that documented their incomes, migration patterns, and other livelihood indicators such as health and education. We might also have usefully devoted more attention to the land market per se. The constitutional change, the wide ranges in commodity prices, and the land-use complications arising from the extended drought interacted in ways to greatly complicate the number and prices of land sales. Farmers were not eager to talk about these sales, and our study failed to provide a consistent tracking of land values. This area of research remains important, however, particularly as the Yaqui Valley agricultural sector continues to consolidate.

A final, and very important, limitation of our study is that we did not consistently have a policy advisor on the ground as we worked through our scientific questions. That is, we did not hire someone with the relevant knowledge and skills in economic development, politics, and policy analysis who could live in the Yaqui Valley and help promote the recommendations from our research. We were extremely fortunate to have a leading agronomist and a water expert as part of our team, who could work directly with producers to improve input use efficiency and reduce environmental losses from agriculture. We also conducted a study of knowledge systems (chap. 5) that described how scientific knowledge is passed on to producers. However, we lacked a person who could play a *normative* role on the project; that is, individuals whose roles were to provide policy options to decision makers in Hermasillo and Mexico City, and to help outline the types of economic incentives needed to diversify and stimulate agriculture wisely in the Yaqui Valley.

Despite these limitations, four conclusions seem to dominate our sixteen-year analysis of the valley; we suspect that similar conclusions would apply to many other intensive agricultural regions throughout the world. First, globalization of the valley was not scale neutral. Meeting international competition proved especially hard for small commercial farmers and the *ejido* sector as they lost much of their protection via special policies. As a consequence the valley became more efficient, but probably less equitable.

Second, it is easy to talk about diversification into high-valued agricultural products, but very difficult to implement such a change—even in regions with good infrastructure and substantial human capital. The knowledge systems for these products are very different than for staple food commodities (chap. 5). More generally, the switch in thinking from a supply-oriented, producer-dominated approach to a demand-driven, consumer-oriented viewpoint is a large hurdle involving new institutions and forms of market organization.

Third, as countries develop, macroeconomic policy becomes even more important for agriculture. At the same time, agriculture's direct influence on exchange rates, credit policy, and fiscal policy is diminished. Under these circumstances, agriculture is increasingly buffeted by forces over which it has little control.

Finally, the Yaqui experience underscores the pressure on both policy makers and farmers for dealing with uncertainty and variability. As economies move toward market-determined (versus policy-determined) prices, and toward international pricing, price variability becomes a dominant fact of life. The large price swings in the Yaqui Valley, particularly in the decade of the 1990s, had enormous impacts on farm profits. Farmers, during years of very low prices, were in rather desperate economic straits, which raised difficult political-economy questions of whether or not the government should intercede with varying kinds of price and income supports. Whether or how governments should intercede in commodity markets that exhibit substantial price variability strikes us as a large, but unfinished, part of the ongoing debate on globalization.

For all of the foregoing reasons, the economic sustainability of agriculture in the Yaqui Valley remains precarious. With substantial crop supports—at least cyclically—for wheat, maize, and cotton, the incentive to diversify is low, and the need for high input use (water, fertilizers, and pesticides) remains. Moving away from supports is all the more difficult when producers' leading competitors in the United States and Europe receive ample subsidies of various kinds. On the other hand, the potential of rising cereal prices, driven by global meat and ethanol demands, may lessen the need for future subsidies. However these various forces play out, Yaqui farmers will be better off in the long run if they can develop and adopt a more diversified, market-oriented approach as a means of gaining profits and managing risks.

Acknowledgments

We thank Marshall Burke, Arturo Puente-González, Dagoberto Flores, Jessa Lewis, Amy Luers, Pamela Matson, Andrew Melaragno, Ellen McCullough, Justin Kitzes, Josh Goldstein, and Samantha Staley for research contributions to this chapter; Derek Byerlee, John Dixon, Prabhu Pingali, and C. Peter Timmer for critical reviews of the manuscript; and the Ford Foundation, Pew Charitable Trusts, McCaw Foundation, MacArthur Foundation, David and Lucile Packard Foundation, and Stanford University's Freeman-Spogli Institute for International Studies for financial support of the research.

Chapter 9

Agricultural Research and Management at the Field Scale

Ivan Ortiz-Monasterio and David Lobell

Development specialists often distinguish between three stages of agricultural development in a region: expansion of agricultural areas, increased use of inputs, and increased input-use efficiency. The Yaqui Valley was early in the third stage of development when we began our studies. Our intent was to help meet the growing need for improved efficiency as farmers struggled to remain profitable, costs of inputs rose, and irrigation allocations were reduced (chaps. 3 and 8). Equally important, we were interested in managing and reducing the environmental costs of agriculture, and we hoped that advances in efficiency would be one path toward win-win situations for farmers and the environment (chap. 3).

As in most irrigated agricultural systems worldwide, nitrogen (N) and water are the two main inputs to crop production in the Yaqui Valley. Therefore, much of our work has focused on improving field-level efficiencies of N and water use. This included studies of alternative N and water management practices and their impacts on wheat yield and farmer profits. Significant effort has also been devoted to understanding yield responses to inputs other than N and water. These studies were motivated by the recognition that, for example, if a factor other than N was limiting yields, then removing this constraint would increase yields and therefore N use efficiency, since a greater yield would be produced with the same amount of applied N. In addition, we realized that the effects of changing N or water management on yields often depended on other factors, such as weather, planting dates, or soil type. If we wanted to better understand how yields responded to N, and how N could be used most efficiently, we would need to understand the interactive role of other factors as well.

This chapter summarizes our work on crop management, beginning with an overview of the current cropping system and management trends, followed by discussions of our work on N management, water management, and wheat yield constraints in farmers' fields. We conclude with a discussion of lessons learned, opportunities lost, and some recommendations.

The Dynamic Cropping System, 1981–2008

Several surveys conducted throughout our project, as well as prior surveys by the International Maize and Wheat Improvement Center (CIMMYT), have provided knowledge of crop selection, management practices, and farmer characteristics. Table 9.1 presents a summary of average values for some of the variables measured since 1981. While some management practices, such as phosphorus application rates, have remained fairly stable, most aspects of the cropping system changed considerably over the lifetime of the project.

One of the most pronounced changes occurred with the arrival of the silverleaf whitefly (*Bermisia argentifolii*) in 1994–95 (Pacheco 1998). Due to the susceptibility of the soybean plant to attacks by the whitefly, the appearance of this insect resulted in a complete loss of the soybean crop in 1994. As a result, the common wheat-soybean rotation was eliminated after 1997, which was unfortunate since wheat (a winter crop) and soybean (a summer crop) could be grown within a period of twelve months and the biological advantage of planting a nitrogen-fixing legume (soybean) in sequence with a nitrogen-using grass (wheat) was lost (Naylor et al. 2001). Since the loss of the soybean crop, it has been difficult to find a replacement summer crop. However, the fact that there has not been enough water in the reservoirs for the planting of a summer crop since 1994 (except for those farmers with wells) has made this a moot point, at least for the time being.

Another adjustment in the system derived from the lack of water for summer crops was the near elimination of wheat residue burning. When farmers followed the wheat-soybean rotation they had a very narrow window between harvesting wheat and planting the soybean crop. Therefore, what most farmers did was to burn the wheat residue, irrigate and plant soybeans. This allowed them to eliminate the time and cost associated with the tillage operations. When the pressure of planting a summer crop disappeared, farmers began to incorporate the wheat crop residues in the soil. Around the same time the government began to enforce legislation that prohibited the burning of crop residues.

Another relevant change has been the reduction in number of irrigations since the early 1990s. During 1998–2003, reservoir levels reached historic lows and improving water-use efficiency became a central topic. The irrigation district reduced the allocation of water to most farmers from five to four irrigations and considered further reductions to three. These decisions of going from five to four irrigations were based in part on experiments by scientists from the National Institute of Forestry, Agriculture, and Livestock Research (INIFAP) and in part on experimental results from a farmers' club. The latter showed that it was better to eliminate one irrigation and maintain the wheat area than to reduce the wheat area and maintain the number of irrigations.

A major shift in the production from bread wheat to durum wheat also took place during the project's lifetime. In 1991, 100 percent of the wheat area was planted under bread wheat, and in 2001, 80 percent of the wheat area was planted to durum wheat. This change was caused mainly by the disease karnal bunt (durum wheat has better tolerance to karnal bunt than bread wheat) and not by prices (Naylor et al. 2001; chap. 2).

Most of the irrigated wheat systems around the world plant wheat on flat fields and form basins that are flood irrigated. In 1981, 94 percent of the farmers in the valley utilized this method of planting and irrigation. However, by 2003, 98 percent of the farmers had changed to planting on raised beds and using furrow irrigation. This change took place mainly because bed planting allowed the mechanical cultivation of weeds, which is cheaper than the use of herbicides (particularly those for weed grasses), but also because it requires less seed and allows better water management. Other advantages that have been observed are less lodging (which is the permanent displacement of plant stems from the vertical), allowing the band application of N fertilizer when the crop needs it the most, and less incidence of karnal bunt.

There were also changes in weed control practices, with more farmers applying a preplant irrigation to germinate the first generation of weeds, which are controlled mechanically right before planting. The adoption of bed planting and the use of the preplant irrigation were associated with a reduction in the use of herbicides for weed grasses, which are significantly more expensive than broadleaf herbicides.

The adoption of the preplant irrigation also had a critical impact on N use efficiency because of greater N losses prior to planting. Most farmers apply 65 to 75 percent of their total N rate before the preplant irrigation,

TABLE 9.1 Average values from surveys of farmers conducted in the Yaqui Valley (1981–2008).

Year	*1981*	*1982*	*1987*	*1989*	*1991*	*1994*	*1995*	*1998*	*2001*	*2003*	*2008*
# of farmers surveyed	91	74	41	101	63	85	58	50	75	80	88
Fertilization											
nitrogen (kg/ha)	177	194	219	231	221	251	234	244	263	251	254
phosphorous (kg/ha)	50	46	49	47	44	52	47	53	47	51	56
# applications	1.6	1.9	2.2	2.1	2.3	3.1	2.5	2.6	2.8	2.2	2.1
% that use NH3 gas	20	26	44	48	60	84	86	76	91	90	88
% that use urea	46	51	71	69	48	42	34	58	63	78	63
% that use ammonia	54	59	51	49	38	33	17	18	16	8	5
% that use phosphorus	59	55	83	78	62	67	74	88	80	89	82
Wheat management											
# irrigations	4	4.7	5.5	4.8	5.2	5.2	4.9	4.5	4	3.7	4.4
planting durum wheat (%)	19	18	46	44	0	42	59	67	80	88	74
planting bread wheat (%)	81	82	54	56	100	58	41	33	20	12	26
planting on dry surface (%)	60	60	17	18	5	6	2	14	3	4	2
planting on beds (%)	6	7	34	36	59	55	85	88	84	98	85

Type of herbicide used											
broadleaf (%)	66	56	94	70	74	35	86	88	83	90	93
weed grasses (%)	11	21	0	17	15	47	14	8	9	6	7
both (%)	24	23	6	13	11	18	0	4	8	4	0
Previous summer crop											
soy (%)	28	75	78	24	5	84	81	0	0	0	0
none (%)	61	22	12	68	76	6	9	82	96	98	96
Tenure type											
Group *ejido* (%)	21	18	22	21	19	8	17	6	7	4	6
Individual *ejido* (%)	33	37	32	39	36	22	21	23	27	15	17
Private property (%)	39	42	46	38	37	28	40	25	21	39	56
Rented (%)	8	4	0	3	8	41	22	46	45	43	22
Average age	n/a	n/a	n/a	n/a	n/a	n/a	n/a	48	50	53	55
Professional degree (%)	12	12	12	22	24	29	35	29	36	35	54

then they irrigate, and twenty days after irrigation (on average) they plant wheat. During this twenty-day period, important losses of N occur (chaps. 3 and 10). Experiments in farmers' fields receiving a preplant irrigation demonstrated that applying 225 kg N/ha one day before planting resulted in a 50 percent N use efficiency compared to a 41 percent N use efficiency by applying N about twenty days before planting (Ortiz-Monasterio, unpublished data).

Starting around 1994, farmers began to increase the size of their operation through renting so that farming could remain financially viable. There has been an increase in the number of farmers with professional training, and in 2003 the average age of the farmer was fifty-three.

Of all of these changes to the cropping system, perhaps none have been more relevant to our project than the increase in the rate of N applied in wheat in the Yaqui Valley. The N use during the 1980s was on average 205 kg N/ha; in the 1990s it increased to 238 kg N/ha; and between 2000 and 2008, it had reached 256 kg N/ha. The increase in N use during the 1980s was encouraged by subsidies and probably associated with the adoption of the preplant irrigation and the associated reduction in N use efficiency. The increase in N use during the 1990s might have also been associated with the switch from bread wheat to durum wheat, which tends to yield 5 to 10 percent more than bread wheat in the Yaqui Valley. The average yield for the decade of the 1980s was 4,858 kg/ha, while the average yield during the 1990s increased to 5,218 kg/ha. In addition, the yield potential increases between the 1980s and 90s suggest that the demand for N in the crop, and thus fertilizer inputs, would also increase. A slight increase in average yields was apparent after 2000, but this is mainly attributable to favorable climatic conditions (fig. 9.1).

Improving Nitrogen Use Efficiency and Reducing Nitrogen Losses

Early studies with irrigated spring wheat at the research station demonstrated that using split applications with some of the N applied at planting and most of the N applied at the stage of development of Zadoks 31 (Z31; beginning of stem elongation, this usually takes place about forty-five days after planting) had generally resulted in higher yields than applications of all N at planting or Z31. N applications were more efficient at Z31 than at planting, but it was also shown that some N was needed early in the crop cycle, particularly when the soil was highly N deficient (Ortiz-Monasterio et al. 1994).

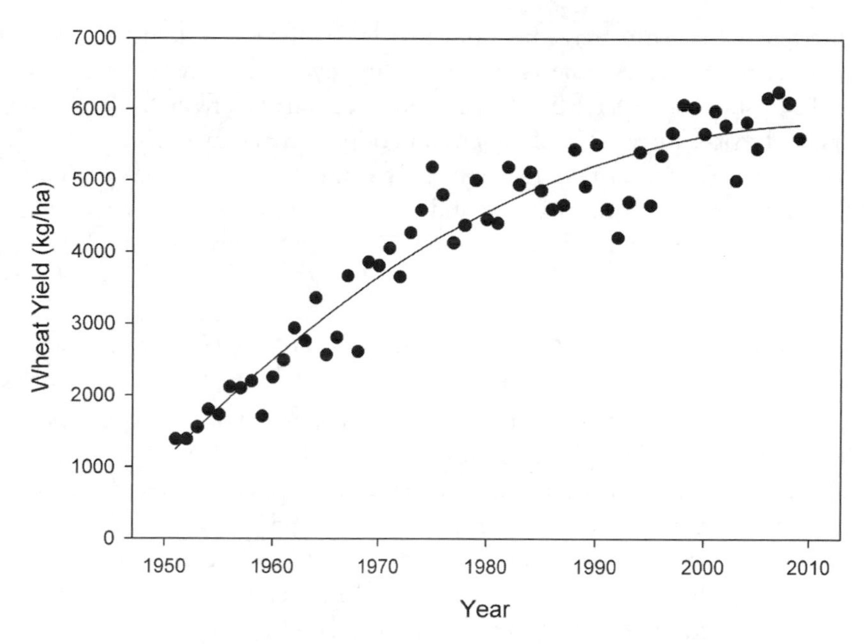

FIGURE 9.1 Average Yaqui Valley wheat yield, 1950–2009 (Stanford/CIMMYT surveys).

Building on this information, we carried out a set of experiments (described more fully in chap. 3) that modified the amount and timing of fertilizer application compared to farmers' practice. The results showed that our "best" alternative N management, which reduced the N rate from 250 to 180 kg N/ha and changed the timing of N application to better match N supply from fertilizer with demand by the plant, reduced losses to the environment and did not affect yield or quality. The robustness of this best alternative practice was tested for two crop cycles (1996–97 and 1997–98) in a total of twenty-eight on-farm evaluations[1] that represented different rotations, soil types, and management practices in the Yaqui Valley.

The results showed that the optimum economic rate, *averaged* over all locations using the same timing as our best alternative, was 194 kg N/ha (wheat price US$0.17/kg; N price US$0.54/kg),[2] supporting the results of the integrated studies at the research station, which showed that 180 was more profitable than 250 kg N/ha.

While the best alternative seemed to be reasonable, subsequent surveys suggested that there was relatively little adoption of the practice. Two potential issues that had come up in earlier surveys seemed to be important. First, changing the timing of the N applications raised the concern that farmers would face an increased risk due to untimely rains that could interfere with their management practices and could result in lower yields. Second, the fact that some years are climatically much better than others, and that some fields have better soils and more nitrogen reserves than others, and that neither are well understood or predictable by farmers beforehand, suggest that farmers may shy away from any practice that prevents the best yields in the optimal year and low N reserve soils. In the following sections, we analyze the issues around both of those concerns.

Logistical Challenges and Their Impact on Timing of Application

The best alternative practice required two changes in the timing and proportion/distribution of N application compared to the farmers' practice. The first change was to eliminate the preplant applications and instead make the first N application at the time of planting. The second change involved applying a larger proportion of the N at the time of the first postplant irrigation (67% instead of 25%). The general guideline for these changes was to better match supply with demand by applying most of the N at the time of rapid demand by the crop, which coincides with the time of the first postplant irrigation (fig. 9.2).

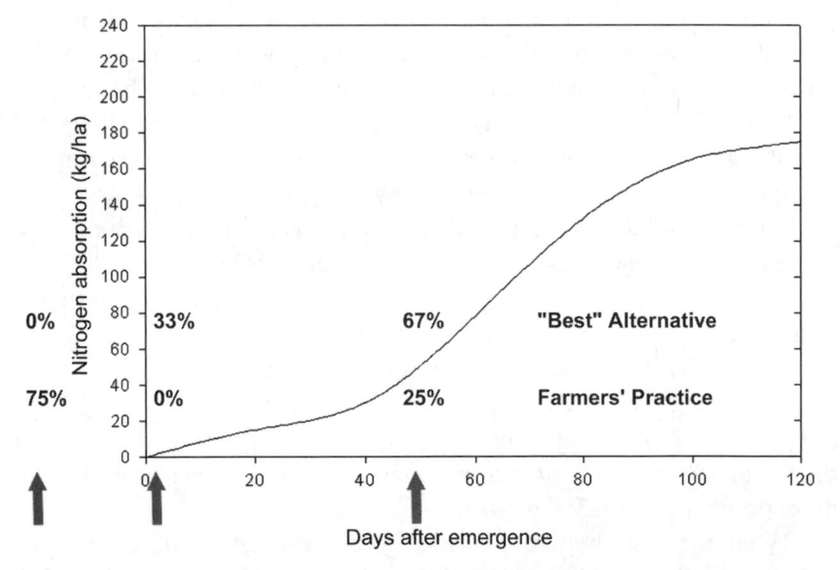

FIGURE 9.2 Cumulative N absorption in spring wheat and the traditional (farmers' practice) vs. the "best" alternative timing for N application. Numbers represent the percentage of the total N rate applied during these three different times.

The on-farm trials demonstrated that the timing we selected for our best alternative, which applies most of the N later in the crop cycle, had a significant increase in N use efficiency at all levels of N applied, except at the very high rate of 300 kg N/ha where the N application rate was so far above the crop demand that regardless of how the N was applied there were high N losses (fig. 9.3). These results further suggested that changes in the timing of the N application would be of benefit to the farmers and the environment.

However, our later surveys of management practices carried out to monitor changes showed very limited adoption of this recommendation in spite of the improvements in efficiency. There were no changes in the rate of N used, and farmers were still applying most of the N preplant. Two factors seemed to be involved in limiting the adoption of the best alternative practice. First, preplant applications occur on flat, dry soils, which allows the use of large equipment that can fertilize fields very rapidly (10–15 ha/hour). The probability of rain in late October and early November (see fig. 9.4), which is when most farmers apply the N preplant, is lower than during the time of planting (mid November to late December). Since farmers

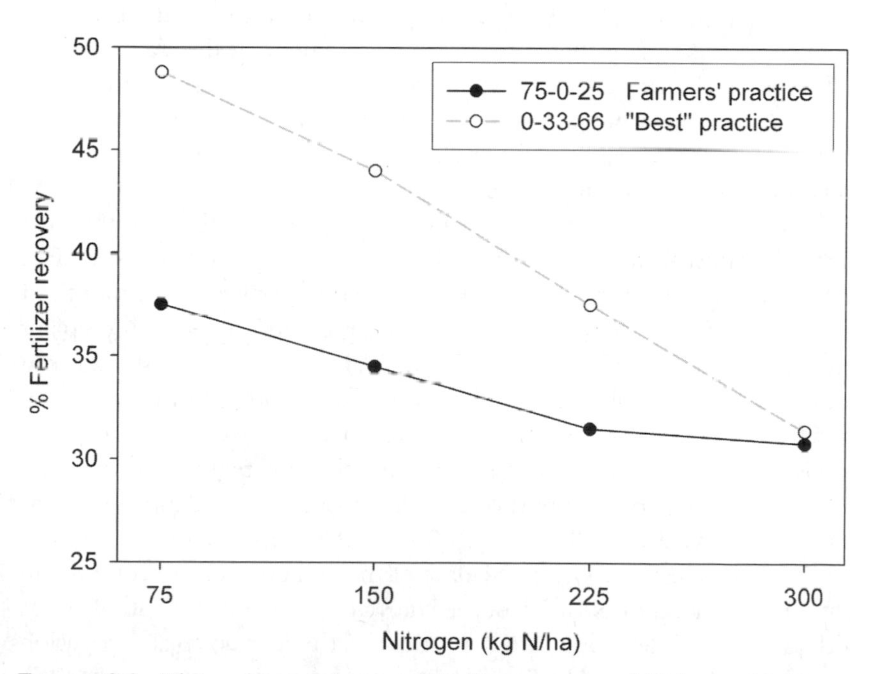

FIGURE 9.3 The average percentage of fertilizer recovery of 28 farmers' fields using different rates and timings during the 1996–97 and 1997–98 crops cycles in the Yaqui Valley.

are using a preplant irrigation, that means that to apply the N close to planting (a day or two before planting) or at planting requires operating in wet soils, which limits the size of the equipment that can be used because of problems with soil compaction and traction. Therefore, the large equipment that is used to apply N in dry soils cannot be used on wet soils, resulting in a slower and more labor-intensive operation for fertilization. In addition, when fertilization and planting take place at the same time, it takes longer to plant than if the farmers only concentrate on planting.

During planting farmers want to get the planting done as soon as possible since there is a risk of rain during the optimum planting date window (November 15 to December 15). If a heavy rain occurs, particularly in early December, this most likely will result in the farmer having to plant outside the optimum planting date. Planting outside the optimum planting date does not necessarily result in lower yield; however it increases the probabilities of lower yield substantially (INIFAP 2001). In some years planting after December 15 can result in a 20 percent yield reduction (Ortiz-Monasterio and Lobell 2007). Planting and fertilizing at the same time may only represent a delay of one or two days compared to only planting, but this can make all the difference in planting on time or late if the rain happens to take place that day. This risk is something that farmers with relatively small fields are more willing to take than large farmers that plant many hectares distributed in several plots across the valley. As mentioned, the trend over the last years has been for farmers to plant larger areas so that they can remain competitive.

The second factor involves having to apply most of the N at the time of the first postplant irrigation, which takes place forty-five or fifty-five days after planting. This means that most N would be applied in the month of January, when the probabilities of rain are significant (fig. 9.4). A preplant irrigation is approximately equivalent to 120 mm of rain. Most fields can be planted between eleven and thirty days after the preplant irrigation (depending on soil texture). Taking an average of about twenty days for most fields in the Yaqui Valley, and assuming that the time required before driving heavy machinery is proportional to the amount of water applied, means that if we have a rainfall event of 30 mm, the farmer can move in with equipment to fertilize within five days of the rain event. If we consider an intermediate level of N deficiency in the soil, for each day that the farmer delays the N application around the time of the first postplant irrigation the yield will be reduced by 55 kg/day, which after five days represents 275 kg/ha of wheat (Ortiz-Monasterio 2002). On the other hand, that delay in N application also represents an increment of 0.25 percent units (e.g., 11%

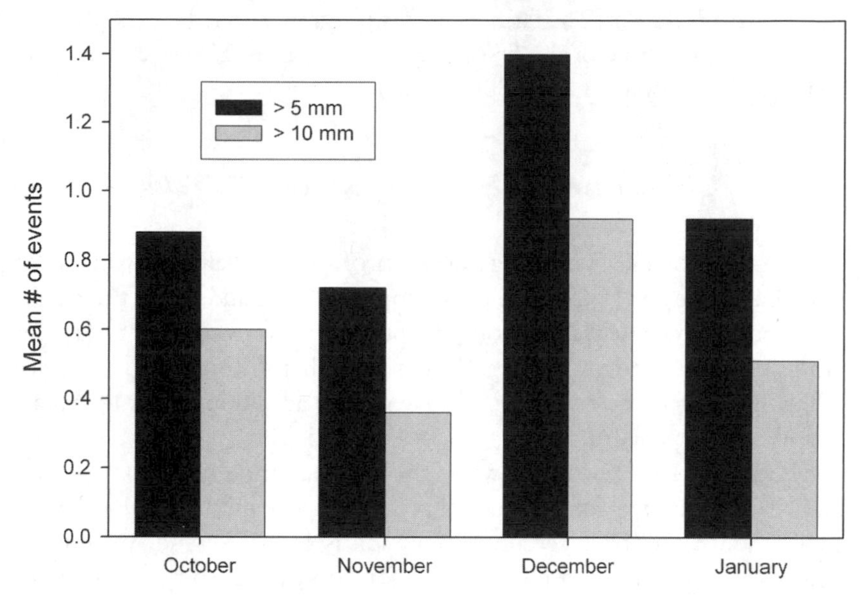

FIGURE 9.4 The average number of days with more than 5 mm or 10 mm of rain, October through January, 1980–2003.

to 11.25%) of protein in the grain, which represents better quality, since N applied late tends to improve protein. The probability of a 30 mm rain event in January is less than one in five years. The level of risk that a farmer is willing to take can vary widely; for some this is an acceptable risk, for others it is not. Again, farmers with large holdings who have a number of plots distributed around the valley will face more problems trying to apply most of the N at the time of the first postplant irrigation because of logistical problems (personal communication from farmers).

The on-farm trials showed that if we maintain the N rate constant and only modify the timing to better match supply with demand, there are improvements in both grain yield and quality. However, the impact in quality is larger and more consistent than in yield. This would suggest that a program that is willing to pay a price premium for high protein wheat could represent enough incentive for farmers to change the timing of the N application. There is an early indication that this may be the case. In the 1995–96 crop cycle, a grain company was interested in buying high-protein wheat in the Yaqui Valley. This company contracted directly with farmers and offered a price premium of US$17.1[3] per ton if the protein was above 13 percent. The farmers in this program applied less N preplant and more at the time of the first postplant irrigation. However, they also increased their total N rate to 300 or even 350 kg N/ha.

Uncertainty about How Much Nitrogen Is Needed

A second factor likely affecting adoption of our best alternative was the spatial and temporal variations in N supply and demand. After the results of our integrated studies came out, the price of nitrogen fertilizer rose, but farmers kept applying high rates. We expected that the increase in the price of fertilizer would be a strong incentive for the adoption of our best practice. What happened?

Babcock (1992), working with US Midwest farmers, found that optimal N fertilizer rates for risk-neutral producers may increase if uncertainty about the weather or uncertainty about soil N levels exist. He established that it was optimal for farmers to plan for good years, so that N will not limit potential profits on those years. He showed that uncertainty in the levels of soil N resulted in a 25 percent increase in the N rate applied by farmers. In addition, he reported that when growing season uncertainty and soil N uncertainty were present, optimal N application increased by 36 percent in his simulations. This meant that because of the inability of farmers to forecast the weather or reliably estimate the amount of residual soil N, farmers were fertilizing for the good years and the poor soils.

In the Yaqui Valley, average wheat yields have varied between roughly 4–6 t/ha over the last two decades, mainly due to changes in temperature and solar radiation (Lobell et al. 2004; Lobell et al. 2005b). These changes in yield potential result in significant changes in N demand by the crop depending on the climatic conditions in a given year.

In addition to climatic variability, substantial differences exist in residual soil N available to crops, in part related to soil type, but also due to previous management (e.g., crop and N fertilizer rate). A set of on-farm trials established in the Yaqui Valley included a zero N treatment, which was used to measure residual soil N by measuring total N uptake in the plant at maturity. The results from twenty-eight locations showed that the average amount of residual soil N was 109 kg N/ha with a range (44 to 201 kg N/ha) and standard deviation (47 kg N/ha) that showed a large degree of variation (fig. 9.5). The optimum N rate for the site with the highest and the lowest amount of residual soil N, assuming this residual amount was

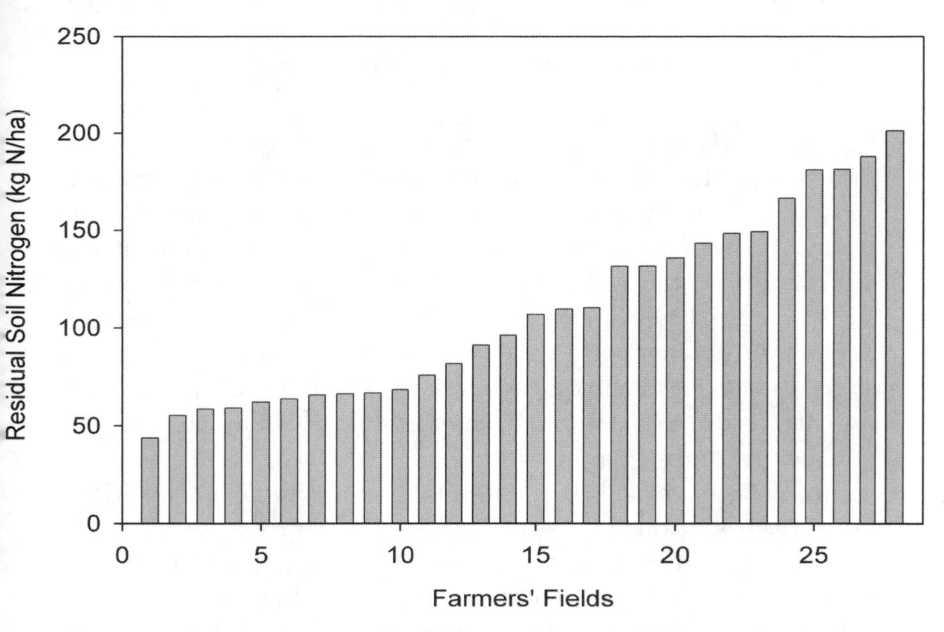

FIGURE 9.5 The amount of residual soil nitrogen in 28 farmers' fields measured during the crop cycles 1996–97 and 1997–98 in the Yaqui Valley.

known, was estimated to vary from 248 to 0 kg N/ha, showing the high degree of variability from one field to the next (Ortiz-Monasterio, unpublished data). To gauge the optimum practice in light of this uncertainty, a study similar to that of Babcock (1992) was done in the Yaqui Valley. An N management decision model was developed for an irrigated wheat system that incorporated hypothetical diagnostics of soil N and growing season climate. The model was then used to quantify the potential value of these forecasts with respect to wheat yields, farmer profits, and excess N application. Under the current situation faced by farmers, uncertainty in soil and climate conditions, the model accounted for an average overapplication of N by roughly 35 percent. Soil variability was found to be roughly three times as important as climate variations in terms of potential impacts in profits in the region (Lobell et al. 2004). The results were remarkably similar to those of Babcock (1992). Both studies, done completely independent of each other and with different types of farmers growing different crops, suggest that farmers are making rational choices when they fertilize in large amounts, given the range of field conditions that they face and when environmental costs are not considered.

Site-specific Nitrogen Recommendations

We began working in 1994 on the development of diagnostic tools to help farmers account for the amount of residual soil N. Soil tests that measured initial levels of soil nitrate as well as incubation studies to measure mineralization rates provided insufficient information (unpublished data). A number of plant diagnostic tests used in other wheat-growing regions of the world (e.g., basal stem nitrate concentration, total N in the plant and chlorophyll concentrations) were evaluated in the valley, but while these diagnostics provided an accurate measure of N needs in any individual year, they tended to be affected by the yield potential of the crop and were therefore not consistent from year to year.

In 1997, in collaboration with Oklahoma State University (OSU), work began evaluating the use of a sensor-based approach for site-specific N management in irrigated wheat in the Yaqui Valley. The N sensor sends a beam of light in the red and infrared bands into the wheat canopy and collects reflectance data from the leaves in these two wavelengths. This data is used to calculate the normalized difference vegetation index (NDVI), which is used to predict yield potential. The NDVI is measured in an N-rich strip planted within the area that needs to be diagnosed for N deficiency. The N-rich strip is an area of the field that received a rate of N high enough to guarantee that there will not be N deficiency in the field. The N in this strip has to be applied at the latest by the time of planting. The NDVI values are collected toward the end of tillering and beginning of stem elongation in wheat (Z31). Through the use of a crop algorithm this index is used to predict yield and the need for any additional N (Raun et al. 2005).

The well-fertilized strip allows a visual comparison between the farmers' management and the well-fertilized N strip. A farmer that has experience growing wheat can easily establish if there are differences or not between his field and the well-fertilized strip. If there are no noticeable differences between these two areas for the trained eye, then no additional N is needed. This means that the sensor is not really necessary when there is no N deficiency in the farmer's field with respect to the well-fertilized strip. However, the sensor becomes very useful when the area under farmers' management shows a deficiency since the sensor will help identify the optimum N rate. Although technically the sensor is not necessary when there are no visual differences between the well-fertilized strip and the farmer's field, we have noticed that farmers still like to collect the readings with the sensor. For them, collecting the data with the sensor when there are no visual differences is equivalent to getting a second opinion and reassures them in their decision-making process.

Prototypes of the current GreenSeeker sensor were tested in the Yaqui Valley and suggestions were made for their improvement. Experiments were established at the CIMMYT research station and in farmers' fields to generate data sets for the development of a local algorithm for optimal site-specific N recommendations. Important modifications to the original algorithm developed for winter wheat by OSU were suggested and implemented for irrigated spring wheat, largely reflecting the fact that spring wheat can fully recover from N deficiency as late as the Z31 stage (http://nue.okstate.edu/CIMMYT/INSEY_Ciudad_Obregon.htm).

Technology and Knowledge Transfer

We can divide our activities when working with farmers into two phases: validation and technology transfer. The validation work with the GreenSeeker sensor began during the 2002–03 and 2003–04 wheat crop cycle. During those two years a total of thirteen experiments were established in farmers' fields in the Yaqui Valley. These experiments were one to two hectares in size comparing conventional N management used by the farmers versus the use of the GreenSeeker sensor, and they showed that over all locations farmers were able to save 69 kg N/ha on average without any yield reduction. The price of US$0.90 per unit of N in the valley at that time represented US$62/ha of savings to the farmers. These initial results were very encouraging but additional tests evaluating a wider range of conditions and larger plots (10 ha or more) was needed before adoption could be widespread (Ortiz-Monasterio and Raun 2007).

At a first glance one would have expected that the change associated with timing of N applications would have been easier to transfer than the change in rate. However, during the transfer of the sensor technology to farmers they seem more willing to test the sensor than to change the timing of their N application. In fact, in this early stage of technology transfer farmers are maintaining the same timings and proportions of N application. This means that the farmers using the sensor still apply most of the N preplant. Therefore, the potential for saving N is not as large as if they applied a smaller amount earlier in the crop cycle. Basically the use of the sensor so far has consisted in helping the farmer decide if there is a need to apply additional N during the first postplant irrigation or not. As some farmers become more confident with the reliability of the sensor as a diagnostic tool for N recommendations, they seem more willing to start changing the timing and distribution of their N applications. They are realizing that in order to exploit the maximum potential savings with the use of the sensor they need to apply less N basal. Some farmers have expressed interest in applying zero N either preplant or at planting and letting the sensor tell them how much N to apply at the time of the first postplant irrigation. This would allow them to maximize their potential savings.

An extreme example of the maximum savings that farmers could obtain is the example of a farmer that applied 275 kg N/ha by the time of planting. This farmer left an area of his field with zero N application. Readings with the sensor at Z31 did not show any difference between the area with zero N and the rest of the field that received 275 kg N/ha, suggesting that no N was needed (presumably because of very high residual nitrogen remaining from previous years' fertilizations). At harvest this was confirmed by measuring yield in both areas, which showed no difference in yield.

There are at least four other examples similar to the one just described where there was no difference in yield between a zero N application and the farmers' traditional N rate, which was always above 200 kg N/ha. This is consistent with the results of the residual levels of soil N in our on-farm trials (fig. 9.5), which identified a number of farmers who had such high levels of N in the soil that no N was needed at all. On the other hand, there are farmers who would have severe N deficiencies if they did not apply any N, again pointing to the importance of site-specific N management.

The potential saving in N fertilizer through the use of the sensor will first increase as farmers are willing to reduce their basal N applications and take advantage of the N stored in the soil. However, these gains will likely be reduced through time as the levels of residual soil N are depleted. Nevertheless, the continued used of the sensor should still be more profitable and more environmentally friendly than the current practice.

Some of our earlier work identified the important role that the farmers' unions play in the decision-making process for N recommendations (chaps. 3 and 5). That meant that we had to convince not only the farmers but also the technical departments of the farmers' unions, which are responsible for ensuring that the farmers are using appropriate technology. Therefore, we decided to work with the farmers' unions' technical department as well as with the farmers. To do this we approached PIEAES (Agricultural Research and Experimentation Board of the State of Sonora), an old CIMMYT partner, who represents all farmers in the state of Sonora on issues related to agricultural research. Through PIEAES, we made contact with AOASS (Association of Producer Organizations of Southern Sonora), which represents several farmers' unions. A technology transfer proposal was written to the state agricultural research funding body (Fundación Produce Sonora) requesting funds for the purchase of eight of the GreenSeeker sensors. This proposal indicated that if the proposal was approved, the farmers' unions would cover 50 percent of the cost of each of these sensors.

PIEAES was very instrumental in lobbying and getting the consensus from the farmers' unions to agree to invest in these sensors. This was very important since having invested some of their resources in the purchase of the sensors the unions felt a sense of ownership in the transfer of this new technology. We have been working on technology transfer for six years; the average N savings during this period of time has been 70 kg N/ha. During the crop cycle 2008–09, there were approximately 7,000 hectares using this technology. We are also working on transferring this technology to other areas of intensive agriculture around Mexico and Asia.

We have already taken steps toward improving N recommendations from an average N rate to a specific recommendation for each farmer's field (see section on technology transfer). We have also identified weather forecasts as an important component for improving N use efficiency. The next step was to address within-field variation to try to further increase N use efficiency. Within-field yield variation has been shown to be present in Yaqui Valley wheat fields using satellites that have a resolution of 30 × 30 m (Lobell et al. 2003). We have been testing in farmers' fields a GreenSeeker multisensor system mounted on a sprayer and connected to a variable rate application system. This system was specially developed for conditions in the Yaqui Valley using one of the sprayers locally available in the market and adapting it to the bed planting system to apply UAN32[4] in the furrows using drop nozzles. This allows us to diagnose N needs and address variation at the 10 m^2 level in real time, on the go. Early results are showing that N rates applied using the multisensor system compared to the handheld sensor were not significantly different. Considering that the multisensor technology is at least six times more expensive than the handheld and technically more challenging, we do not foresee adoption in the near future. However, as the rate of N is reduced in the Yaqui Valley to improve efficiencies, it is likely that the N variation problems will start to increase. Therefore, the use of new cutting-edge technology may be justified. Economic analysis that considers the cost of the equipment versus the potential saving for applying N rate at the 10 m^2 scale will have to be done to establish if this technology will be profitable to farmers in the region.

One would expect that it is a matter of time before this type of technology is adopted. The N to wheat price ratio versus the cost of the new sensing technology will be an important factor in determining how soon adoption may take place. On the other hand, prototypes of a "pocket sensor" that will cost around US$100, and that will also measure NDVI, will be tested in farmers' fields in the 2008–09 crop cycle. The low cost of this new technology will allow direct access to many farmers in the Yaqui Valley, changing the dynamics of the technology transfer effort.

Water-Use Efficiency

The water-use efficiency in farmers' fields is a major concern throughout irrigated systems worldwide, especially given the growing scarcity of water in many regions. As discussed in the section on the dynamic cropping system, improved efficiency of on-farm water use became a major concern during the recent drought. However, decisions on water allocations were based on only a few experiments performed by farmer groups and INIFAP.

While efficiencies at the regional scale had been investigated throughout our project with reservoir management and module water-allocation models (see chap. 11), the potential impact of reducing per-area allocations on wheat yields and total water-use efficiency had not been studied in detail. In response to the perceived need for more scientific information in water-allocation decisions, we conducted an analysis of field irrigation management and water-use efficiency. Using a previously validated crop simulation model, we showed that the impact of reducing water allocations was strongly dependent on the year and location of the crop (Lobell and Ortiz-Monasterio 2006).

For example, significant rainfall occurs during some but not all growing seasons, and the amount of this rainfall influences wheat yields for fields with two to three postplant irrigations (fig. 9.6). The initial amount of water stored in the soil, down to 90 cm depth, also had a major impact on whether and by how much reducing irrigations would impact yields. Unfortunately, reliable data on typical values of initial moisture in farmers' fields were not available, and therefore firm recommendations on irrigation amounts could not be made. However, the study allowed us to identify the limitations of experimental results upon which the current policies were based, as well as highlight the specific data needed to better answer questions about optimal irrigation strategies.

By the time the study was completed in 2005, reservoir levels had rebounded and the perceived need to improve field water-use efficiency had

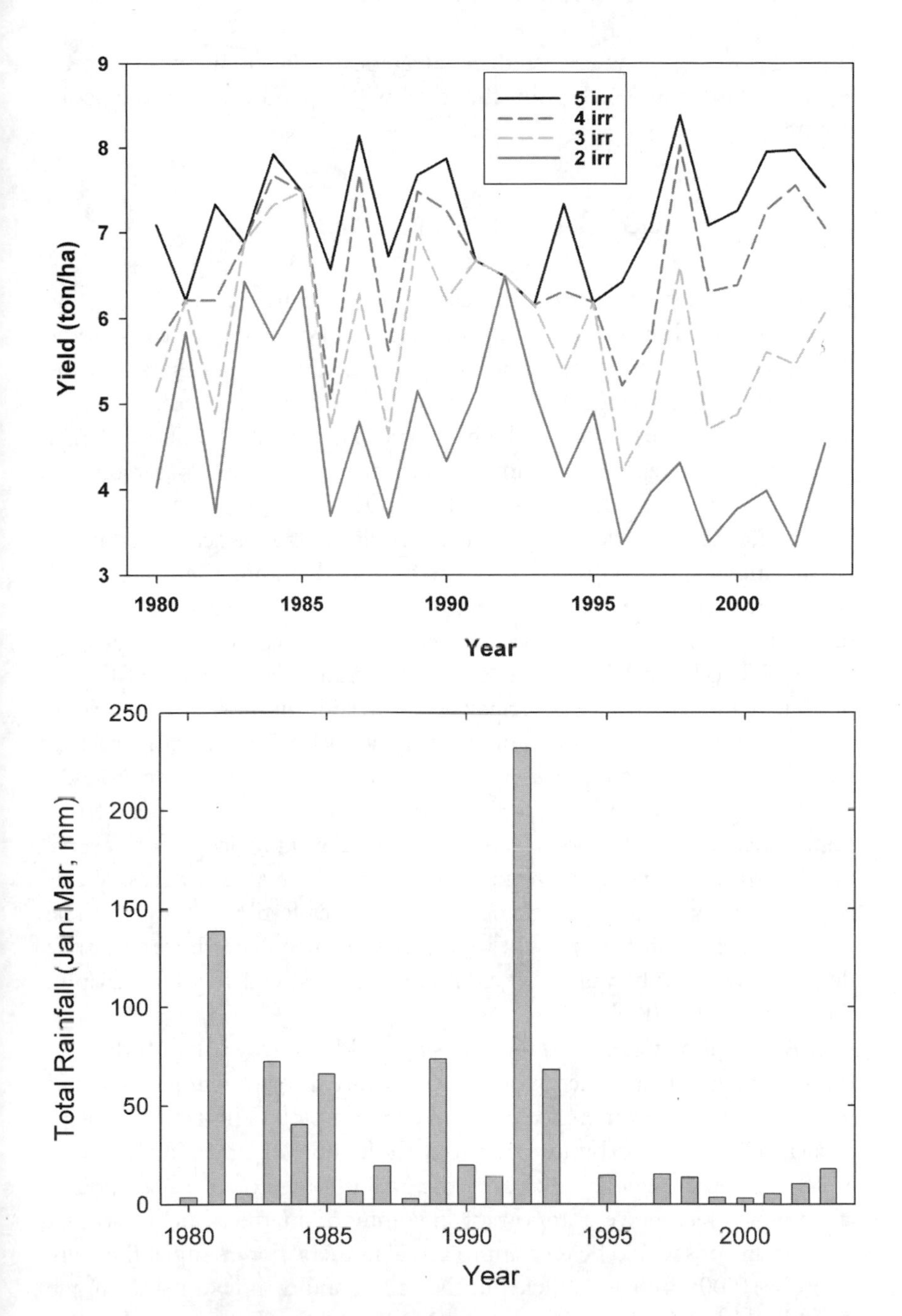

FIGURE 9.6 Modeled yields for 2–5 irrigations by year (top). The simulated yield loss for reduced irrigations depended on the amount of rainfall during January–March (bottom), with irrigations making a smaller difference in years with more rainfall (Lobell and Ortiz-Monasterio 2006).

faded. However, the issue of reduced irrigations will surely revisit the valley in the future, at which point the study will hopefully inform the policy debate.

Yields Constraints in Farmers' Fields

Despite a widespread desire to increase productivity and profits, growers often have difficulty achieving high yields. For example, while some growers in the Yaqui Valley report yields of 8 t/ha or more, others get only half as much. On average, yields in the valley are roughly 25 to 35 percent lower than those achieved on the highest yielding fields. Similar or larger gaps between average and maximum yields are seen in every major agricultural region in the world (Cassman et al. 2003).

Achieving yields close to genetic potential is difficult because so many factors interact to control crop growth, and therefore there are many chances for something to go wrong. Weather variables, such as temperature, rainfall, and solar radiation, can be important, as can several properties of soils on which the crops are grown. Added to these are a host of management factors, such as irrigation and fertilization practices, crop variety selection, planting dates, and planting densities. Biotic factors, such as crop diseases and weed or pest infestations can also play important roles.

Throughout this project, we have attempted to understand the relative importance of these factors in farmers' fields. This knowledge could serve as a basis for identifying management changes that would increase yields or maintain yields while reducing inputs such as fertilizer or water. The focus has been primarily on wheat, as it has been and continues to be the dominant crop in the valley. Wheat is also the most widely grown crop in the world (FAO 2006b).

A logical first step to understanding yield controls is to conduct an experiment in which a factor of interest is varied for different plots, and to then evaluate the difference in yields between plots. This traditional approach has been used hundreds, if not thousands, of times in the Yaqui Valley and other regions. However, the goal of our project was to understand which factors were important in commercial farmers' fields, so that our conclusions would be relevant to actual farmers. For example, there are roughly 10,000 individual fields in the valley, and each field has a unique set of biophysical and management factors that differ from each other and the experimental station. Our objective was to determine which factors represented the most critical constraints to yields on the greatest number of fields, and which factors were relatively unimportant.

Ideally, we could have conducted experiments on a large number of farmers' fields in multiple years. However, the resources required to do this were beyond the scope of our project. In addition, many of the factors that could affect wheat yields (such as weeds, diseases, or various soil properties) are difficult or impossible to manipulate experimentally. We therefore relied on a statistical, or empirical, approach where yields were measured in many fields and compared with measurements of factors that were thought as possibly important.

Meeting Data Needs

This type of approach invariably requires data on yields, climate, soils, and management, which are often hard to obtain in agricultural regions such as Yaqui Valley. A significant amount of effort was therefore devoted throughout the project to obtaining reliable datasets on yields, climate, soils, and management. In some cases, reliable datasets already existed and were acquired through local contacts. For example, records of regional wheat area and production since 1960 was provided by SAGARPA (Ministry of Agriculture, Livestock, Rural Development, Fisheries, and Food) and daily weather records since 1969 were available from CIANO (Center for Agricultural Research in the Northwest). In other cases, new datasets were created through surveys, field work, and remote sensing data analysis. Landsat satellite images, for example, proved to be the most reliable source for measuring crop yields for individual farmers fields across multiple years (fig. 9.7; Lobell et al. 2003). Often the data served not only to answer predetermined questions, but to generate new questions based on viewing and discussing interesting patterns. For example, visualizing the degree of spatial variability, such as shown for yields in figure 9.7, often spawned questions about the relevance of experimental trials in a particular location, or about the importance of a previously unconsidered factor.

Temporal Yield Variability

As mentioned, time series of regional yields and daily meteorological data were among the first datasets available in this region. Using these, the effects of weather on average wheat yields were investigated, and surprisingly strong relationships were uncovered. Average regional yields since 1980 were negatively related to average nighttime temperatures during the growing season (fig. 9.8). Higher temperatures likely reduce yields by

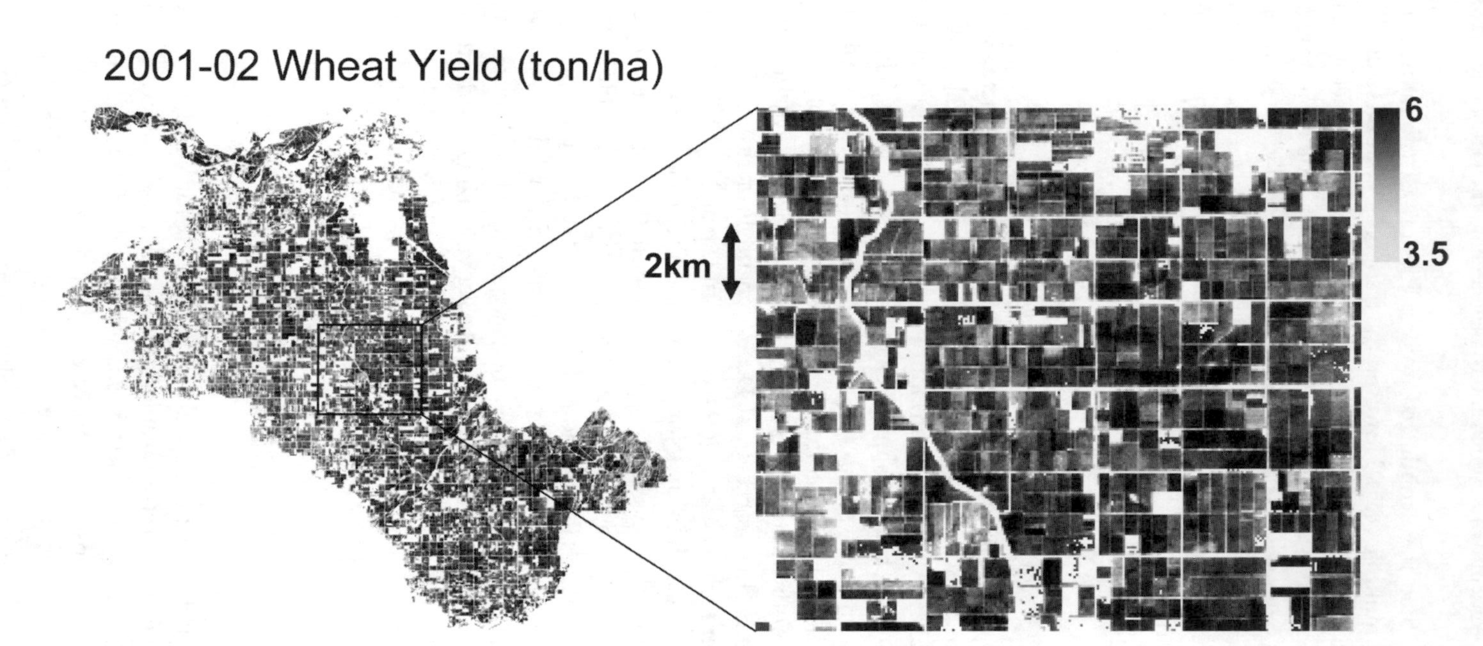

Figure 9.7 An image of wheat yields for the 2001–02 growing season derived using Landsat satellite data, showing large variations between different fields. Similar images were developed for each growing season since 2000 and used to evaluate causes of yield losses in farmers' fields.

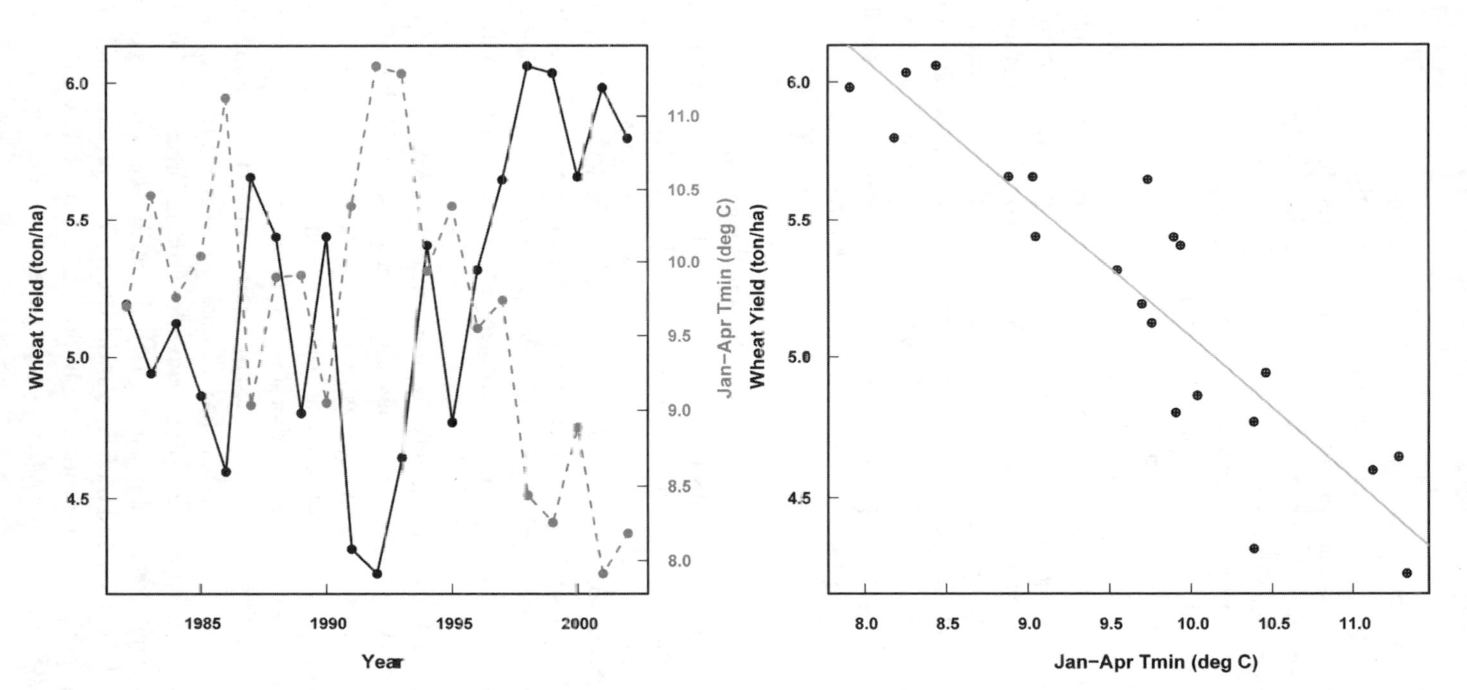

FIGURE 9.8 Time series of wheat yields (ton/ha) and January–April average minimum temperatures (left). Years with cool growing season nights tend to produce high yields (right, $r^2 = 0.81$).

increasing development rates and thereby shortening the length of time the crops have to grow and accumulate grain. The reason that nighttime temperatures (rather than daytime or average temperatures) figured so prominently was not entirely clear but appeared related to the fact that years with high daytime temperatures tend to also have higher amounts of solar radiation, which promote photosynthesis rates and cancel the negative temperature effects (Lobell and Ortiz-Monasterio 2007).

The fact that such a high fraction of yield variation could be explained by temperature had at least three implications of interest. First, other factors such as the introduction of new varieties, increased fertilizer applications, or other management changes since 1980 have apparently had only a small net effect on yields (Lobell et al. 2005b). This conclusion was somewhat surprising, since several factors had changed over time (e.g., increased rental of *ejido* lands, increased N rates) and the time series of yields appeared to show a step increase after 1990 (fig. 9.8). That this jump was simply reflecting a cooling of temperatures in recent years implied that once this cool period ended, yields would revert to pre-1990 levels. In fact, yields did drop after the study was conducted, with yields in 2002–03 of just 5 t/ha.

Second, yields of the major crop in the valley appear quite vulnerable to future climate changes, with a 1°C average increase in temperature expected to reduce yields by roughly 10 percent (Lobell and Ortiz-Monasterio 2007; chap. 6). Climate models project that average temperatures will increase in this region between 0°C and 3°C by 2025 and 1°C and 5°C by 2055 (IPCC 2007c). Coping with climate change is therefore a substantial challenge for the Yaqui Valley, and opportunities to improve resilience, for example by developing more heat-tolerant wheat varieties, should be explored.

Third, an accurate model of regional yields opens up many opportunities for growers and others involved in the valley's wheat industry. For growers, if forecasts of growing season temperatures can be made in advance of planting, they could adjust factors such as planting date or N rates based on their expectation of whether it will be a cool, high-yielding year or a warm, low-yielding year. As discussed in the previous section, model simulations showed that rational, profit-maximizing farmers would reduce N rates in most years if they were provided with a reliable forecast of climate, thereby reducing excess N application (Lobell et al. 2004). Forecasts could also be used by farmers and traders to anticipate regional production levels and better negotiate pricing contracts. Importantly, a model based only on January and February temperatures does essentially as well as the model in figure 9.8, and therefore predictions are available at least two months before harvest even if based on measured and not predicted weather.

Spatial Yield Variability

While giving insight into temperature controls, regional scale data offered little hope for uncovering how factors such as soil or management affect yields, since these factors are fairly constant in time when averaged over the entire region. To understand nonclimate effects on yields, we developed maps of yields at 30 m resolution from satellite remote sensing data (Lobell et al. 2003). Surveys of farmer practices, data on soil properties and land tenure, and other remote sensing products were then compared with yields for several years to test various hypotheses about what factors contribute to yield losses in farmers' fields. One of the first questions we asked was whether the productivity on *ejido* lands, which make up about half of the valley, were systematically different than those on privately owned lands. In contrast to the expectations of many, no significant differences were found. This may be explained by the high rate (up to 80% in some years) of renting out *ejido* land to private farmers, although even yields in fields that we knew to be managed by *ejidatarios* were not systematically lower.

An analysis of 2001 yield estimates with management surveys suggested that the factor that best explained yield differences between farmers was N application rates (Lobell et al. 2005a). This was a surprise given the previous experimental data and analyses that suggested farmers apply much more N than needed by crops. However, 2001 was an unusually cool year with record high yields. As a result, crops required more N to reach their climatic yield potential.

When the analysis was repeated in 2003, the most important factor was the timing of first postplant irrigation, with delayed irrigation leading to reduced yields. This year had lower yield potential than 2001, and thus N rates were less important. In addition, uncertainty about water availability because of reduced reservoir levels caused many farmers to delay irrigations beyond typical dates for normal years.

These analyses showed that a few management factors could explain a large fraction of yield variability in any single year, meaning that most farmers were faced with similar yield constraints. However, the results also highlighted the year-dependence of any analysis of crop yields or associated quantities such as N use efficiency. The optimal management practices change from year to year, and so farmers must make decisions whose effects on yields and profits cannot be predicted accurately without knowing what the weather will be. Since no factors consistently reduced yields in both years, there were no easy recommendations for improving yields. Similar studies would have to be carried out for a period of five to ten years to see whether certain factors are more commonly associated with high yields.

Additional studies focused on individual factors for which other datasets were available. For example, remote sensing provided reliable maps of planting dates on individual fields, as well as the density of weeds in the summer prior to planting. Comparison of these maps with yields revealed a significant but strong year-dependent loss of yields due to planting dates outside the officially recommended interval of November 15 to December 15.

Wheat yields in fields with weeds prior to planting were also evaluated using summer images of weed densities, with yields on weedy fields averaging roughly 5 percent lower than weed-free fields in each of two years of analysis (fig. 9.9). Calculations showed that the loss of profit from a 5 percent yield loss is more than the cost of weed control by two diskings, and thus suggested that better weed control is a sensible option for improving yields (Ortiz-Monasterio and Lobell 2007).

Other studies considered the role of soil type and climate in spatial yield variation, indicating that each play a statistically significant but relatively minor role (Lobell et al. 2002; Ahrens 2009). As with management, the effects of soil properties varied significantly by year. For example, data on soil texture from 181 fields throughout the valley, obtained from a local fertilizer company, were compared to remote sensing estimates of yields for six different growing seasons. In 2003, a year when irrigation practices played an important role, as discussed earlier, soil texture explained roughly 25 percent of yield variations. Thus, the ability of soils to hold water can be quite important in years when water is the main limit to crop yields. However, in the other five years, the correlation between yields and texture were much lower.

Technology Transfer

As discussed in the section on yields, much of the effort for the yield studies involved generating datasets needed to address the project questions. For the remote sensing analysis, this often involved communication with farmers to confirm that our interpretations of satellite data, in terms of wheat yields, planting dates, or some other variable, were valid. In this process, growers not only helped to refine our estimates or identify new

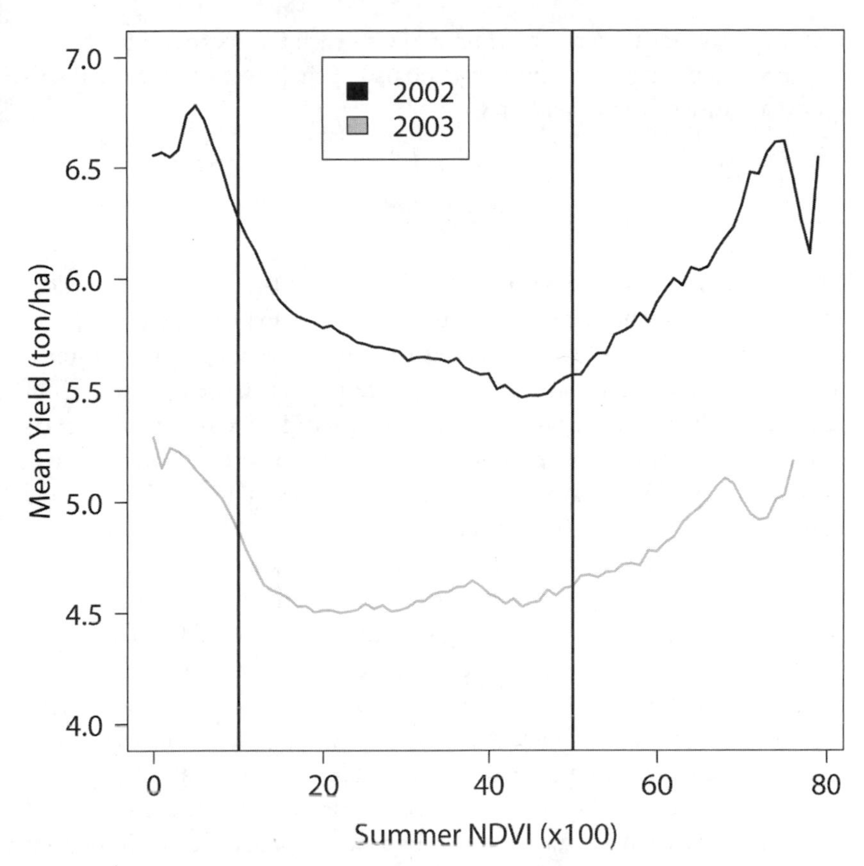

FIGURE 9.9 The average wheat yield from remote sensing vs. a measure of vegetation abundance (NDVI) in the previous summer. Low and high NDVI values correspond to bare and cropped fields, respectively, while intermediate values (between the vertical bars) indicate the presence of weeds. Reduction of yields at intermediate NDVI suggests that weeds were associated with up to 1 t/ha loss of yields on average for each of two study years (Ortiz-Monasterio and Lobell 2007).

questions, but they often were very keen to use the satellite data for their own purposes. For example, one grower mentioned that he used the yield images to identify a low-yielding part of his field, and after confirmation in the field, decided to apply herbicide to this subarea.

To further promote the transfer of yield images and their applications, we established working relationships with several farmers who managed a total of twenty fields, where they received images of their fields and provided feedback on our datasets. Unfortunately, we have found that obtaining feedback on specific applications of the images was difficult without spending a significant amount of time to visit each farmer individually. The PIEAES is allowing and supporting the use of their website for the distribution and use of these yield images.

Implications

What lasting lessons have been learned from these studies on wheat yields? The lack of any management factor in consistently explaining a significant amount of regional yield loss indicates that there are not one or two main management practices that will have a large impact in yield increase, but rather a number of management practices that will have marginal impacts on yield. Changes in consistent factors such as farmer skill, education, risk aversion, or access to capital would not likely result in significant yield changes. Instead, uncertainty about future weather conditions appeared to be the major obstacle to reducing yield losses. This knowledge should help to focus current efforts on generating and using climate forecasts. The effect of late irrigation in 2003 also suggests that institutional factors, such as the delay of water delivery to farmers, can be important.

While soil properties were not critical for high yields in most years, yield differences between fields with good and poor quality soils were greatest in years that were climatically unfavorable (Lobell et al. 2002). This finding indicated that biophysical resources may become more important as climate changes in the future, and thus that soil properties are important in defining the vulnerability of farmers to poor climatic conditions (Luers et al. 2003).

From a research perspective, the yield studies led to a greater appreciation of spatial and temporal heterogeneity that extended to studies of other aspects of the agricultural sector. When evaluating farmer decision making, we found it useful to first question the relevance of data we had, which was often from the experimental station or for regional averages, to the farmers' particular biophysical settings. For example, we found that wheat yield variability through time was 50 percent greater for individual farmers than for regional averages (Lobell and Ortiz-Monasterio 2007). When considering whether farmers could diversify into other high-value crops, such as vegetables or citrus, we were therefore careful not to assume that data on regional yield variability of these crops were representative of the perceived risk to farmers, which would be more tuned to variability for their individual farms.

Of course, considering spatial and temporal heterogeneity often makes finding simple management or policy recommendations more difficult. It requires more data and often reduces the strength of effects observed in an individual field or year. However, without an appreciation of heterogeneity we cannot develop win-win management recommendations where both the farmer and the environment benefit. In fact, one of the most important recommendations to come out of our project is the need to measure heterogeneity and manage in a site- and year-specific manner.

Lessons Learned and Opportunities Lost

We highlight here some of the most salient lessons learned from our work on field-level management, including what we might have done differently. There were often substantial challenges with obtaining existing datasets from local institutions and evaluating their quality. We also realized late in the project that much of the data we collected, such as in the on-farm N trials, would have been useful for studies of spatial heterogeneity or as validation data for remote sensing algorithms, if only the GPS coordinates had also been collected. GPS data are relatively cheap and easy to measure, yet were often not collected toward the beginning of the project.

Our experience also made clear the importance of working with individual farmers. However, by the end of the project we realized that for technology transfer it was more efficient to engage with and "sell" the idea of the new technology not only to the individual farmer but also to the union that supplies him with credit, because for some farmers the credit union has to approve these new technologies before they can be adopted (chaps. 3 and 5). Also, if the new technology proves to be useful, the union helps in catalyzing the transfer of this technology among the rest of the members of that union.

It was also important to choose unions that were receptive to new technologies, which is not always the case. One interesting example was when we bought and distributed eight sensors to different institutions that had expressed interest in using the sensor. Six of the institutions paid 50 percent of the cost of the sensor. One of the two institutions that did not pay for the sensor received it as a gift for political reasons, and the other never paid the 50 percent. The two institutions that did not pay 50 percent of the cost of the sensor never used it.

As discussed throughout the chapter, some of the main lessons learned concerned the impact of heterogeneity. Heterogeneity was not only important for extrapolating results from the research station into farmers' fields but also for designing and evaluating the on-farm trials. Farmers typically like to apply treatments to large areas to be convinced that a particular practice is feasible at a commercial scale. However, yields are typically measured for a strip within the treatment. In the sensor trials, the farmer treatment strip that was evaluated for yield in some fields was 90 to 100 m from the sensor strip. In those comparisons the results were not favorable to the sensor, not because the sensor did not perform well, but because of soil differences between the strips. From these early results we learned to have farmers select nearby strips for comparisons to minimize soil differences.

Finally, many of the results of our research could and probably should have been communicated more quickly to institutions and people in the valley. For example, work on water modeling and weed impacts were communicated to leaders of the relevant institutions, but it may have been more productive to organize one or several talks and invite several different organizations that could use the information generated. We were most often limited by the time available to do this type of work.

Recommendations

Several recommendations emerge from our work. Some of these relate to short-term changes that present a win-win situation for farmers and the environment. For example, the collaboration with AOASS (which represents most farmers' unions in southern Sonora) together with SAGARPA, FIRA (Trust Fund for Agriculture), and Fundación Produce Sonora has been an effective alliance for the dissemination of the sensor technology among farmers in the Yaqui Valley. This collaboration should expand and possibly subsidize the use of N diagnostic tools since these are cost effective and very effective at improving N use efficiency. Furthermore, this model of collaboration could be used for the transfer of other technologies such as conservation agriculture, which deals with reduced tillage and residue management.

Another short-term recommendation, especially for farmers that farm relatively small areas, is to eliminate the preplant fertilization and apply their first N application at planting. While more difficult to implement for those who have large farms and their fields distributed across the valley, the adoption of larger equipment for planting and fertilizing may allow them to speed up the process and save fertilizer, money, and the environment at the same time.

In addition, the remote sensing work suggested that programs for maintaining fallow fields weed free should be encouraged and supported, since this will result in better wheat yields and improved phytosanitary conditions in the valley in a cost-effective way.

There are also several long-term recommendations we feel are appropriate. First and foremost, we feel that management approaches that address spatial and temporal heterogeneity should be expanded. For example, both experimental and modeling work indicated that site- and year-specific management of N and water hold great potential to improve efficiencies. More work in this area should be pursued, including not only the development of new tools and practices but the evaluation of whether economic incentives to adopt these technologies are justified. In addition, more years of farm management surveys combined with satellite estimated yields would help to further identify the management practices associated with higher yields in the valley. Finally, to improve water-use efficiency in fields, better information on spatially varying soil properties, in particular their water-holding characteristics, should be gathered throughout the valley.

Second, given the clear sensitivity of the wheat crop to increased temperature, and the inevitability of climate change in this region, we suggest that the development of heat-tolerant wheat varieties as well as the diversification of the cropping systems to include more heat-tolerant crops should be a priority for the valley.

Overall, we feel the dual needs of expanding the use of site-specific management techniques and adapting the cropping system to climate change are the two most pressing recommendations that arise from our studies and are critical for food production, economic well-being, and the environment.

Acknowledgments

We are grateful to Pat Wall, Ken Sayre, and Pamela Matson for providing thoughtful comments on earlier versions of this chapter. Dagoberto Flores, Dolores Vazquez, and Maria Elena Cardenas provided critical research assistance.

Chapter 10

Nitrogen in the Yaqui Valley: Sources, Transfers, and Consequences

TOBY AHRENS, JOHN HARRISON,
MICHAEL BEMAN, PETER JEWETT, AND
PAMELA MATSON

In nitrogen fertilizer management, as in so many other ways, the Yaqui Valley represents a microcosm for the study of changes that are happening all over the world. Over the last thirty years, policies and agronomic advances in the Yaqui Valley have led to the development of a thriving wheat industry dependent on large inputs of nitrogen (N) and water. The adoption of high N application rates in the valley reflects trends in many regions of the world. Globally, the inputs of N to agriculture, together with additional N resulting from the planting of legume crops, add more N annually than is fixed by the biological nitrogen fixation in all natural terrestrial ecosystems (fig. 10.1). These enhanced additions of N have been in many cases critical to increased crop yield, but they also have enormous negative consequences for the global system. Because N is such a mobile element, it tends to not stay where it is put; thus additions of N through fertilizer use are linked with losses of N from soils to freshwater and marine systems, and of greenhouse gases and air pollutants to the atmosphere.

When we began our study, the consequences of fertilizer management for emission of trace gases and nitrate losses in developing world agricultural systems had not yet been evaluated and was one of several outstanding questions related to global atmospheric change as defined by the international global change community (the International Geosphere-Biosphere Program) (Matson and Ojima 1990). Moreover, there were very

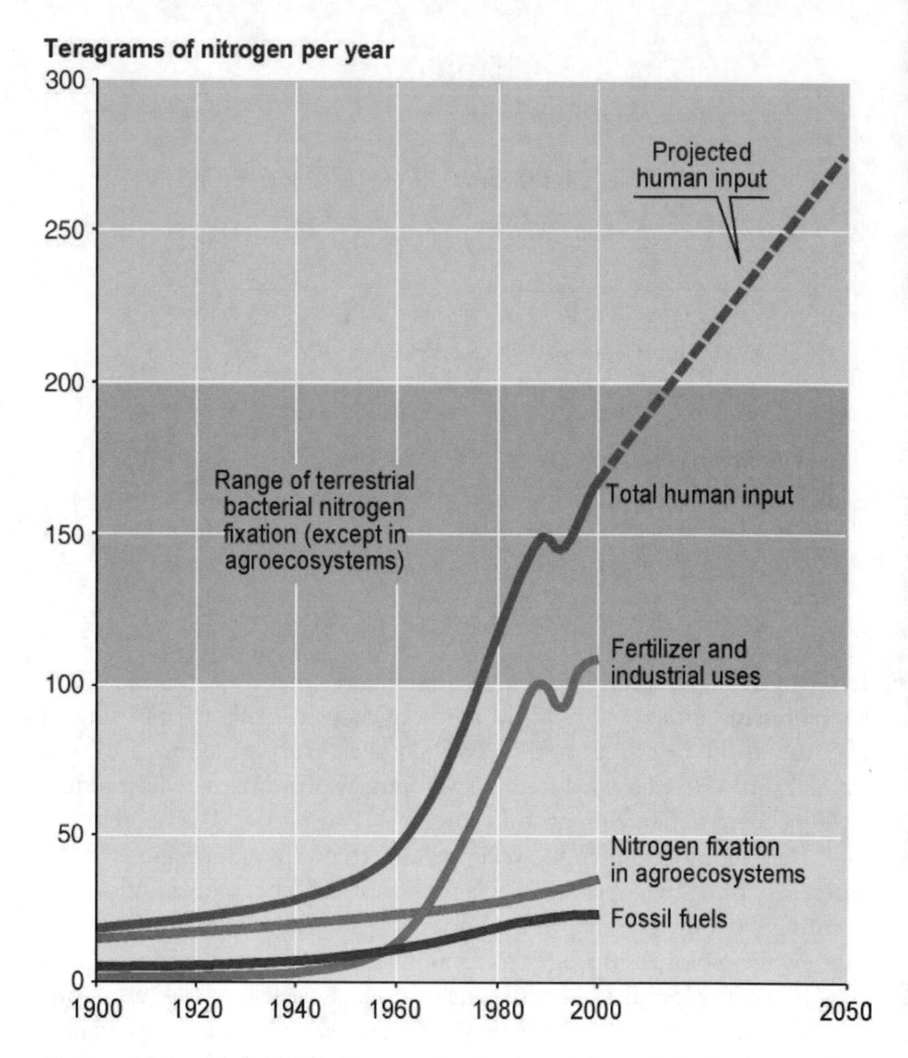

FIGURE 10.1 Global N fertilizer use has increased dramatically in the past forty years, and demand in the near future is expected to continue to grow (Millennium Ecosystem Assessment 2005).

few examples of studies that traced the landscape-scale or watershed-scale consequences of fertilizer inputs. From a purely biogeochemical point of view, these were interesting topics. However, we were also motivated by the growing dialogue focused on sustainability—reconciling the needs for food with the needs for the environment (Conway 1997; Naylor and Matson 1993). We were concerned about the welfare of the farmers and their crops in the Yaqui Valley, but we were also aware of human health issues on land and the threats to the biodiversity in the Gulf of California related to agriculture.

Early in our project, it became clear that fertilizer management in Yaqui Valley fields was relatively inefficient, leading to losses to the atmosphere at rates higher than had ever been measured before, and losses to water systems as well. As we followed N downstream, we found that ecosystems were also reacting strongly to pulses of N from anthropogenic sources. This brought up a new set of questions: When was aqueous N being lost from agricultural fields, and how was it being transformed as it made its way to the coast? Were there negative effects on human health and natural ecosystems downstream or downwind? And how important are agricultural sources of N compared to nonagricultural sources? Many questions still remain unanswered, but our evolving research interests and priorities have led to basic contributions to the field of biogeochemistry at both site and landscape scales, along with contributions to improved management.

Nitrogen Cycling in Agroecosystems

A brief overview of N cycling in agricultural landscapes may be helpful before discussing our efforts to understand N dynamics in the Yaqui Valley. Most N in fertilizers is derived from industrial nitrogen fixation, an energy-intensive process[1] that converts relatively inert atmospheric N_2 gas to forms that crops can use. Worldwide, approximately half of the fertilizer N applied in any given field remains in the crop or soil (Matson et al. 1997). The remaining N fertilizer can take on many forms, with various consequences for ecosystems and public health, before it is ultimately converted back to N_2 (the conversion of inorganic N forms to N_2 is called *denitrification*; fig. 10.2). Some N is emitted from agricultural systems in the form of nitric oxide (NO), a chemically reactive gas that can lead to fine particulate and tropospheric ozone production or can be redeposited as nitrite (NO_2^-) or nitrate (NO_3^-) in ecosystems downwind (see fig. 4.2 in chap. 4). Some N is volatilized in the form of ammonia (NH_3), which quickly goes into solution and is rained out downwind. Some is lost as nitrous oxide (N_2O), which is a stable, extremely potent greenhouse gas. And some is lost in hydrologic flows, as dissolved NO_3^- and NH_4^+ (together known as dissolved inorganic N; DIN), dissolved organic N (DON), or in particulate forms (PN); to a degree, all these N forms make their way to aquifers, rivers, lakes, and coastal marine systems. Finally, some of the nitrogen that is

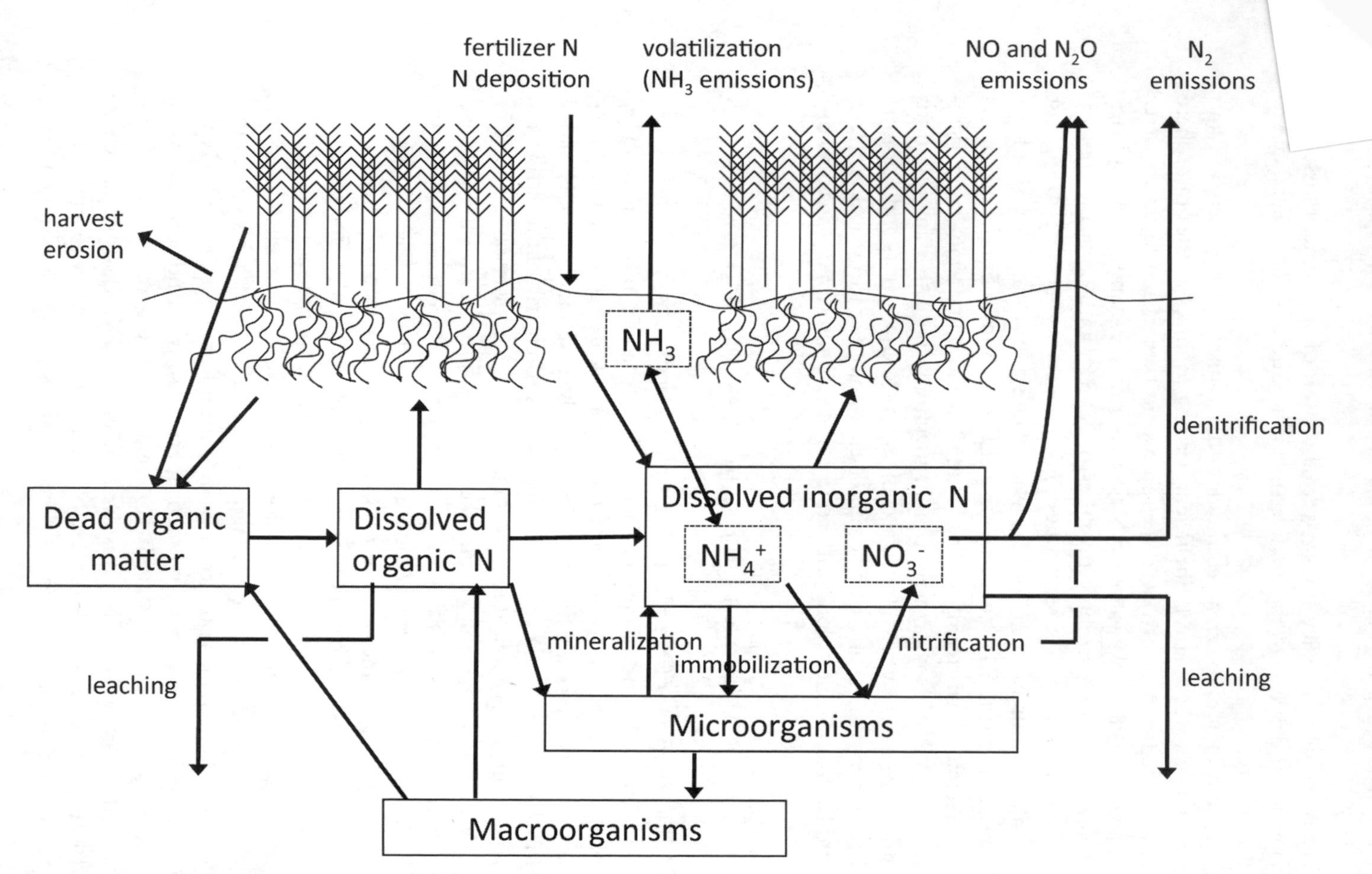

FIGURE 10.2 Simplified terrestrial nitrogen cycle in agricultural systems.

retained in crops flows in trade, through food and feed, and is ultimately concentrated in point sources that can affect water and the atmosphere.

N transported downstream and downwind can act as a fertilizer in natural ecosystems as it does in agroecosystems, with concomitant changes in net primary production, biogeochemical cycling, and species diversity of those ecosystems (Vitousek et al. 1997). In coastal systems, the effects of excess N inputs include changes in phytoplankton species composition, alteration and loss of sea grass habitats, and increases in algal biomass production, among others (Anderson et al. 2002; Carpenter et al. 1998; Diaz 2001; Nixon 1995; NRC 2000; Rabalais 2002). A rapid increase in biomass, or an algal "bloom," can take different forms and have serious effects. Harmful Algal Blooms (HABs), often called "red tides" or "brown tides," can be toxic to fish, marine mammals, and even humans. The decay of algal blooms may also result in hypoxia (low oxygen concentrations) or anoxia (zero oxygen concentrations) in large areas of the coastal ocean (Richardson and Jørgensen 1996); a well-known example is the "dead zone" in the Gulf of Mexico (Turner and Rabalais 1994). These coastal hypoxic zones appear to be increasing in frequency, duration, and spatial extent due to human activity (Diaz and Rosenberg 1995). In 2003, they were identified by the United Nations Environment Programme to be an emerging global environmental threat (http://www.unep.org/geo/yearbook/).

Direct public health concerns associated with N include respiratory illnesses (primarily due to inhalation of ozone, NO_x and fine particulates), pollen allergies (pollen production is stimulated by increased N availability), heart disease (also due to fine particulates), and several problems associated with NO_3^--contaminated drinking water, such as cancer, reproduction and "blue-baby syndrome," or methemoglobinemia (Townsend et al. 2003).

What happens to N in the Yaqui Valley is clearly of ecological and societal importance for residents of the valley and communities both downwind and downstream. Furthermore, our studies in the Yaqui Valley may also offer insights into the fate of N in agricultural systems throughout the developing world in the subtropics—a portion of the world that is critical for current and future global food supply but was severely underrepresented in the biogeochemical literature when our project began (Matson et al. 1999).

In the following sections, we review what is known about transformations and transfers of nitrogen in fields, drainage canals, estuaries, and other sectors of the valley. Our estimates of overall flows are summarized in Figure 10.3, and estimates and sources are provided in Table 10.1.

In the Fields

Farmers in the Yaqui Valley apply ~260 kg N to each hectare of wheat under typical management practices.[2] Only about one-third of the N applied to wheat in the valley ends up in harvested crops, and chapter 9 has discussed crop uptake of N and efforts in the valley to increase N use efficiency by crops. Over the past few decades of research in the Yaqui Valley, we have tried to determine what happens to the other two-thirds. Our earliest efforts in the field were to evaluate management alternatives in terms of their ability to reduce N losses and yet be agronomically feasible and economically attractive (chap. 9).

Using daily to weekly sampling frequencies, we measured changes in soil nutrients, soil gas fluxes, and nitrate leaching prior to and following fertilizer additions in farmers' fields and also in experimental plots (in a randomized block design with four block replicates), where fertilizer additions simulated farmer practice in comparison to alternatives. This biogeochemical research involved the use of gas chromatography, chemoluminescence detection, lysimetry, isotopic labeling, and many other lab and field analytical approaches, and ultimately resulted in the analysis of more than ten thousand gas and soil samples.

Our research on trace gas fluxes and nitrate leaching in the valley indicated that very large losses of N in the form of NO and N_2O (and probably N_2 and NH_3) occur during the preplanting period (Matson et al. 1998; see fig. 10.3). More than 10 kg/ha of fertilizer N can be lost in the form of NO and N_2O during the six-month wheat season, and emission rates measured following the preplant fertilization were among the highest ever reported (Matson et al. 1998; see fig. 13.3). We concluded that, if fluxes of N_2O and NO in other developing world-intensive agricultural systems are similarly high, agriculture is likely to be a more important source of atmospheric change than currently thought, and it will become even more critical in the future. We evaluated several alternative management practices that added less N fertilizer and/or added it later in the crop cycle, in closer synchrony with plant uptake. While all of our alternatives resulted in some reduction in losses, the "best" alternative,[3] which applied 70 kg less N per hectare and applied it during and after planting, lost less trace N gas over the entire cycle than the simulated farmer's treatment lost in just the first month. In addition, neither crop yield nor grain quality decreased, and budgetary savings from reduced fertilizer costs for farmers were equivalent to 12–17 percent of after-tax profits from wheat farming (chap. 3; Matson et al. 1998).

Another interesting finding from our work on trace gas emissions resulted from an isotopic tracer study designed to measure which soil processes were primarily responsible for N_2O emissions.[4] Concentrations of N_2O were measured using gas chromatography, and the ^{15}N content of the gas was measured using mass spectroscopy, allowing us to identify the source of the gas as either nitrification (from an ammonium addition) or denitrification (from a nitrate addition) (Panek et al. 2000). The results of this work suggested that denitrification played a critical role in N_2O and N_2 losses in the period immediately following irrigation or in periods following rains (Panek et al. 2000). Our results also indicated, however, that at lower levels of soil moisture, nitrification was the more important source of N_2O as well as of NO. Thus, contrary to the N cycle descriptions still found in many texts and papers, losses of the greenhouse gas N_2O can occur even when soils are dry, through the aerobic, microbially mediated nitrification process.

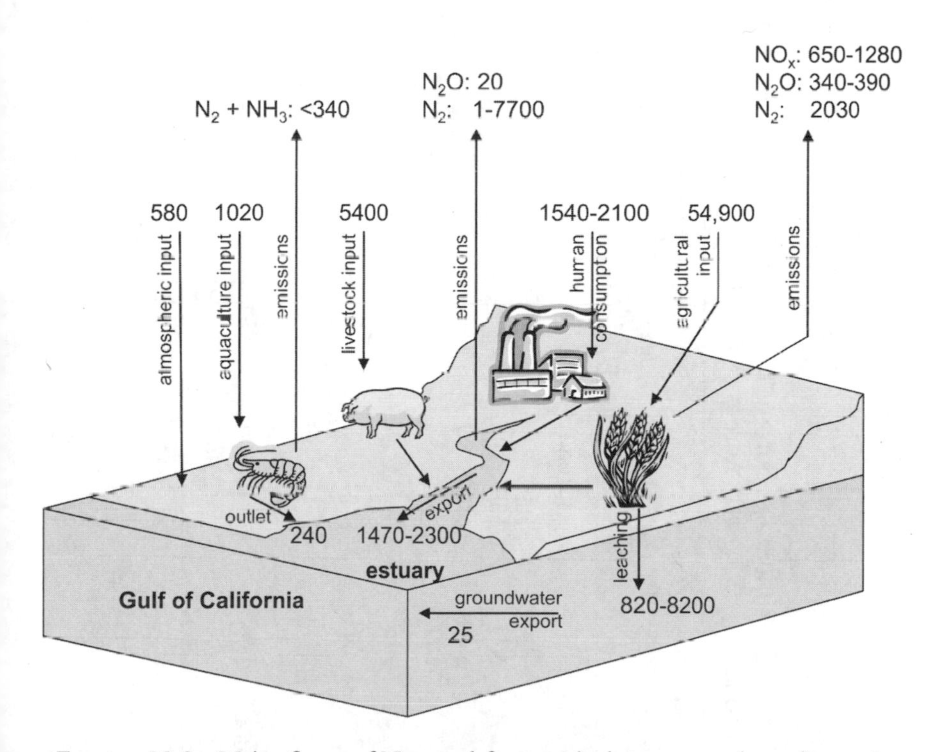

FIGURE 10.3 Major fluxes of N to and from agriculture, aquaculture, livestock production, and human consumption in the Yaqui Valley. Units for all numbers are reported in Mg N/yr. Sources used to estimate each flux are given in table 10.1, except where noted (Ahrens et al. 2008).

TABLE 10.1 Nitrogen inputs and outputs from the Yaqui Valley for the most recent year available by sector (1 Mg N = 1 ton N = 10^3 kg N) (Ahrens et al. 2008).

		Agriculture (Mg N)	*Livestock (hogs and cattle) (Mg N)*	*Shrimp (Mg N)*	*Human consumption/ sewage (Mg N)*	*Drainage canals (Mg N)*	*Total[1] (Mg N)*
Input	N	54,900[2]	5,400[3]	1,020[4]	1,540–2,100[5]		63,500–64,000
Gaseous	NO_x	650–1,280[6]					650–1,280
	N_2O	340–390[7]	4[8]			20[9]	370–420
	NH_3			< 340[10]			< 340
	N_2	2,030[11]		< 340[12]		0.7–7,700[13]	2,030–10,070
Groundwater loading	NO_3^-	820–8,200[14]					820–8,200
Coastal loading	DIN	< 1,690	< 1,690	< 240[15]	< 1,690	660–1,690[16]	660–1,930
	TN	< 2,060	< 2,060	240[17]	< 2,060	1,230–2,060[18]	1,470–2,300

1. Sum of inputs from all sectors in each row. Atmospheric deposition of inorganic N is not shown in the table but is included in the estimated total input. Deposition was estimated to be 583 Mg N (~2.5 kg N/ha, on the basis of Holland et al.'s (2005) estimate of the sum of wet and dry deposition in sparsely populated regions of the West Coast).

2. The sum of fertilizer inputs to wheat, maize, cotton, safflower, and garbanzo beans in 2006. Official estimates of planted area (www.siap.gob.mx) were multiplied by fertilization rates recommended by the National Institute of Forestry, Agriculture and Livestock Research for each crop except wheat (INIFAP; Ortiz-Monasterio, personal communication). Wheat fertilization rates were based on mean practices used by farmers surveyed in the valley in 2003 (Lobell et al. 2005a).

3. Livestock production estimates from 2005 (www.siap.gob.mx) were multiplied by live weight feed-to-biomass conversion ratios of 8:1 for cattle and 4:1 for pigs (Smil 2000) using carcass-to-live weight conversion ratios of 1.61:1 and 1.35:1 (Smil 2002), respectively.

4. N inputs to shrimp ponds in the neighboring state of Sinaloa (Páez-Osuna et al. 1997) were multiplied by the 8,800 hectares of active shrimp farms along the southern Sonora coast in 2004 (www.siap.gob.mx).

5. The valley-wide population of 362,000 in 2000 was multiplied by a per-person intake estimate of 70–100 g protein per day (Smil 2002) assuming 16 percent N content of protein (e.g., Socolow 1999).

6. Measured rates of NO_x emissions (Matson et al. 1998) were multiplied by the area planted to wheat in 2006 (www.siap.gob.mx).

7. Measured rates of N_2O emissions (Matson et al. 1998) were multiplied by the area planted to wheat in 2006 (www.siap.gob.mx).

8. Emissions from manure management in anaerobic lagoons were based on equations recommended by the 2001 Corrigendum of the 1996 IPCC Good Practice Guidance and Uncertainty Management in National Greenhouse Gas Inventories (http://www.ipcc-nggip.iges.or.jp/public/gp/english/).

9. Harrison (2003).

10. Páez-Osuna et al. (1997) estimated shrimp aquaculture ponds in the neighboring state of Sinaloa lost 39 kg N/ha through volatilization or denitrification, equivalent to 341 Mg N for southern Sonora's 8,800 ha in active production in 2004 (www.siap.gob.mx).

11. N_2 emission rates (Panek et al. 2000) were multiplied by the area plated to wheat in 2006 (www.siap.gob.mx).

12. See note 10.

13. Harrison (2003).

14. The proportion of N leached through the soil profile in farmers' fields and at the CIMMYT research station (Riley et al. 2001) was scaled to the entire valley using mean fertilization practices reported by farmers (Lobell et al. 2005) with the area planted to wheat in 2006 (www.siap.gob.mx).

15. The difference between N in inlet and outlet water was estimated by Páez-Osuna et al. (1997) to be 26.8 kg N/ha, equivalent to 236 Mg N for southern Sonora's 8800 ha in active production in 2004 (http://www.siap.gob.mx).

16. Harrison (2003).

17. See note 15.

18. Harrison (2003).

Another potential avenue for gaseous N loss that we did not measure in this set of studies is NH_3, which is released when NH_4^+ is converted to NH_3 in high pH conditions. The pH of soils and irrigation water in the valley are in a range where significant emissions of NH_3 are possible (pH >8 is common), and qualitative measurements suggest emissions could be large. Unfortunately, reliable per-area emission rates of NH_3 are notoriously difficult to measure, and although NH_3 emissions may represent an important flux, they remain unquantified.

Nitrogen not taken up by crops or released in gaseous forms can either be retained in the soil or leached (transported through the soil below the rooting zone in aqueous forms). We found that approximately 20 percent of N is retained in association with soils. Often, N retention in soils is highly correlated with organic matter because of negatively charged surfaces that bond with oppositely charged NH_4^+. Given the relatively low concentrations of soil organic matter in these soils (ca. 0.7%) (Meisner et al. 1992) and a relatively high capacity of the soil to supply N to crops (109 kg N/ha) (Ortiz-Monasterio, unpublished data) the high rates of N retention were somewhat surprising.

To investigate this further, we examined the capacity of Yaqui Valley soils to trap NH_4^+ ions in clay interlayers, making NH_4^+ unavailable for biological uptake until it's released (a mechanism known as "clay fixation"[5]). Our preliminary soil analysis indicates that Yaqui Valley soils contain high levels of vermiculite and/or montmorillonite, both clay minerals that have been associated with NH_4^+ fixation in other regions. Fixed NH_4^+ in such clay minerals have accounted for from 5 to 40 percent of the total nitrogen in the soil (Brady and Weil 1996). Through laboratory experiments, we found that Yaqui soils could fix from 10 to 50 percent of added N in a typical preplanting fertilization, and large amounts of NH_4^+ (140–320 kg NH_4–N/ha) were associated with the fixed pool in the upper 15 cm of soil prior to any N additions in the laboratory (Ahrens, unpublished data). We are currently trying to understand whether NH_4^+ fixed just after fertilization may become available later in the wheat cycle.

Even with higher-than-expected retention rates of N in the soil, significant amounts of N are also leached below the rooting zone (fig. 10.3). When NH_4^+ is nitrified (converted to NO_3^- through a series of biochemical reactions) by soil microorganisms, negatively charged NO_3^- ions move easily through the soil profile when N demand by plants and soil (micro) biota is low. Experiments in farmer fields and under controlled conditions at the International Maize and Wheat Improvement Center (CIMMYT) found variable amounts of leached N, with high leaching rates in farmer fields (14–26 percent of applied fertilizer) and lower rates at CIMMYT (2–5 percent) (Riley et al. 2001). The differences may have been due to soil hydrologic properties or other conditions in the soil prior to fertilizer application (such as the amount of residual N already in the soil prior to the experiments). Even at the lower leaching rates, the total amount of N leached through the soil is quite large. The same best practices identified in the trace gas experiments were able to reduce these leaching losses by 60 to 95 percent.

The large amounts of N transported through the soil profile raised concerns about the fate of leached N. Many communities in the Yaqui Valley are reliant on well water as their primary drinking water source, especially residents living outside of Ciudad Obregón, and well water is commonly used to supplement allotments of irrigation water delivered in surface canals. Unfortunately, we still do not have good measurements of N dissolved in groundwater in the valley. Two efforts to measure the concentrations of N in wells supplementing irrigation water in dry years found generally low levels of both NH_4^+ and NO_3^-, with all but a few samples below the 10 ppm (parts per million) US Environmental Protection Agency drinking water standard. Several samples exceeding the drinking water standard suggest that portions of the Yaqui aquifer may be NO_3^- contaminated, although we know little about either the extent or temporal variability of contamination. This represents an area ripe for further study, with clear implications for human health.

Through the Canals

Not all irrigation water ends up in the valley's aquifer. Rather, some tumbles directly into drainage canals as irrigation tailwater or is transported laterally in subsurface flows to the drainage canals or coastal waters. Water entering the drainage canals from farmer fields then mixes with livestock and urban wastes and is carried toward the coast. Our monitoring and experimental program in the drainage canals was organized around two goals: to understand spatial and temporal patterns of N transport through valley drains and to understand how N is transformed as it makes its way from fields to the coast.

We had two monitoring programs designed to measure the flux of N to the coast in surface waters, with each effort consisting of biweekly sampling spanning two years. The initial program focused for the most part on drains entering Bahía del Tóbari in the southern part of the valley (canals

A1–A7 and C2 in fig. 10.4), and a second effort included sampling from all other major drainage canals in the valley. We monitored multiple forms of N (NO_3^-, NH_4^+, N_2O, PN) as well as a number of environmental variables likely to covary with, or control, N fluxes. We also performed a number of N process measurements and experimental manipulations in whole stream reaches, intact cores, and sediment slurries to determine factors controlling N transformations and N_2O production under controlled conditions. Trace gas (N_2O, CO_2, and CH_4) concentrations and rates of evolution were measured using gas chromatography, and net denitrification was measured using membrane inlet mass spectrometry.

Several themes quickly emerged from the drain monitoring. Water runoff was generally low in the valley—not unexpected for such an arid region—but the concentrations of dissolved N species in the drainage water, including both NH_4^+ and NO_3^-, were quite high (fig. 10.5; maximum concentrations for NH_4^+ and NO_3^- were 40.8 mg N/L, and 21.1 mg N/L, respectively). The total amount of N transported to the coast did not represent a major fraction of the total amount of N applied to agricultural fields (~4 percent each year; fig. 10.3), but given the high fertilizer application rates in the Yaqui Valley, even this relatively small fraction could have a substantial influence in coastal waters—in particular because it is temporally constrained and occurs in pulses (see below). Finally, our monitoring efforts revealed that rates of N export though drainage canals varied on several timescales, ranging from hour-to-hour to interannual variation.

Even at the smallest timescale measured, hourly variations in N transformations had large and important implications for N transport in the canals. Frequent sampling over the course of twenty-four hours revealed rapid changes in N cycling due to shifts in dissolved O_2 concentrations (plankton blooms pumped O_2 into the water column during the day, but not at night when photosynthesis stops). We estimated that failing to measure nighttime N fluxes would have led to a 17 to 38 percent underestimate of N export from one of the Yaqui Valley watersheds (Harrison et al. 2005). In addition to their implications for downstream material transfer, rapidly changing conditions within drainage canals also had implications for greenhouse gas production. Rates of N_2O production changed significantly over the twenty-four-hour period, with N_2O production ceasing for at least eight hours during the night. If we had only sampled daytime N_2O concentrations, we would have overestimated N_2O flux by about 38 percent. The effects of these diel cycles on seasonal and annual rates of N transport, denitrification, and N_2O production appear to be significant.

Comparison of rates across these different systems suggests that in the Yaqui Valley agricultural fields are the dominant source of N_2O to the

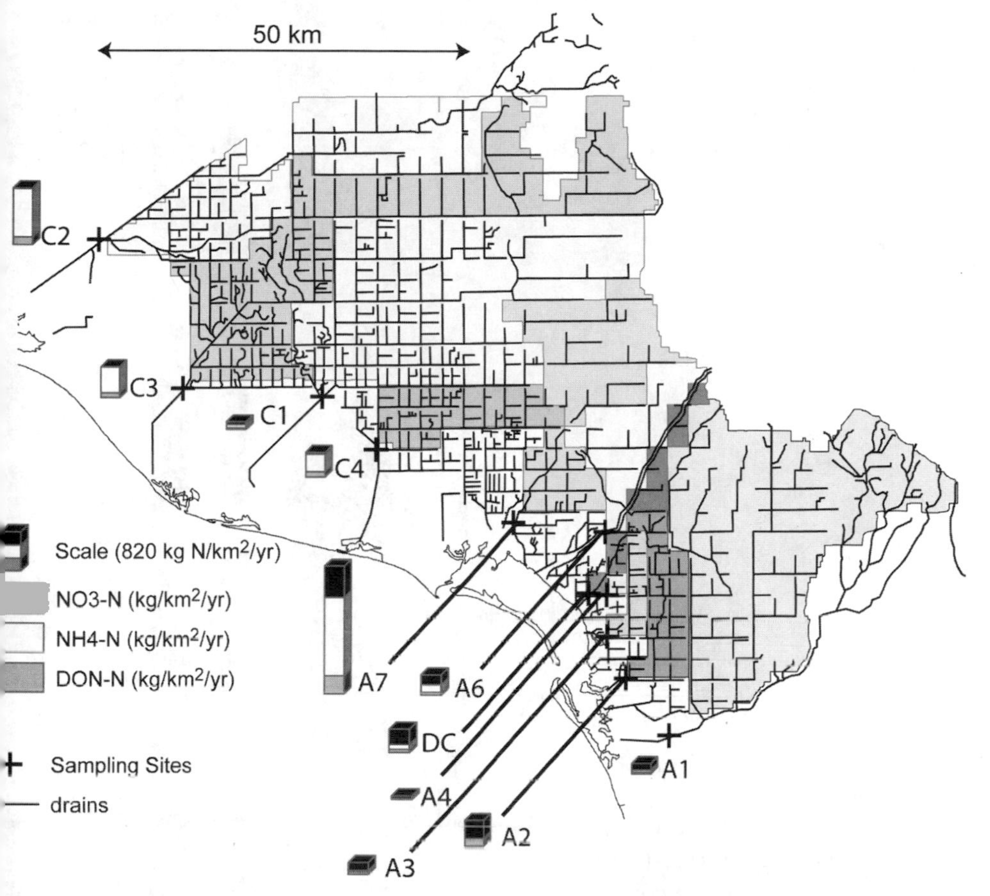

FIGURE 10.4 Mean annual N yields (kg N/km²/yr) for study canals. Height of bars is proportional to N load, and the height of each colored section within a bar is proportional to the contribution from NO_3^-, NH_4^+, and DON for black, white, and grey sections, respectively. Note the spatial variation in N form across the landscape and the dominance of NH_4^+ in the northern watersheds. The dominance of NH_4^+ is unusual compared to other systems where NO_3^- is generally the dominant form of inorganic N in surface waters.

atmosphere. Soils demonstrated both the highest per-area rates of N_2O production and the source of N_2O to the atmosphere with the greatest surface area. Drainage canals also demonstrate high per-area rates of N_2O production, particularly during algae bloom events (Harrison et al. 2005), but because of their small relative surface area, their overall contribution to atmospheric N_2O fluxes is most likely relatively small. The few measurements of per-area N_2O fluxes that we have from estuaries ranged widely,

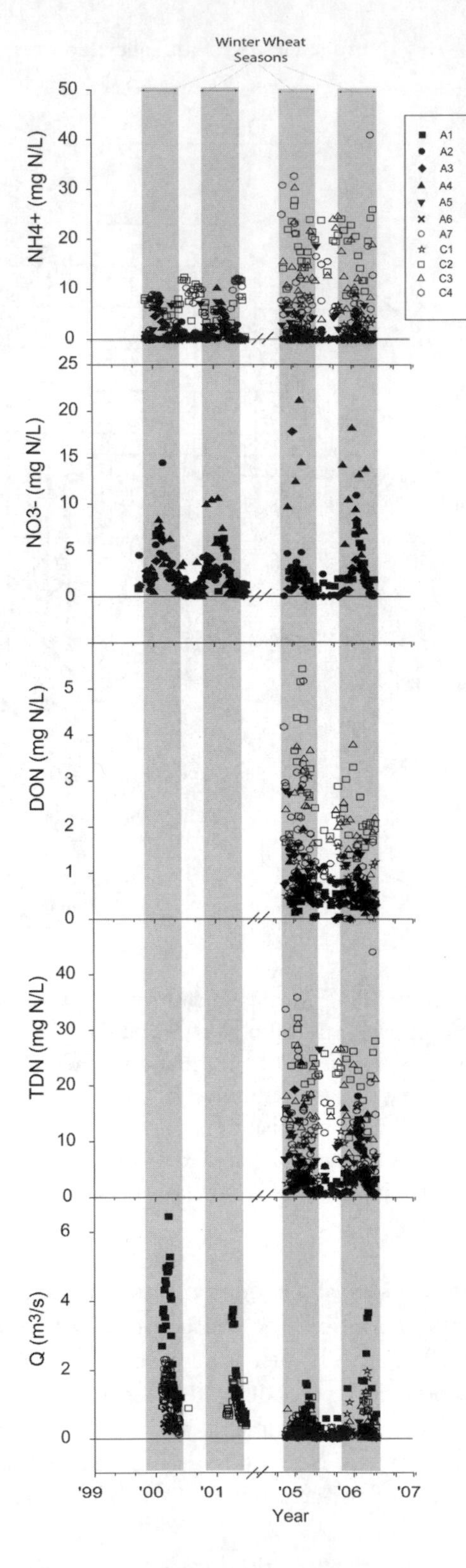

FIGURE 10.5 N forms and water discharge over four years of monitoring. Principal collector drains (C1–C4) and A7 are shown as hollow symbols. Smaller drainage canals receiving water principally from agricultural runoff (canals A1–A6; see fig. 10.4) are represented by solid symbols. DON was not measured during 2000–01. Seasonal variation in coastal N loading is indicated by the elevated concentrations of NO_3^-, NH_4^+, and DON during times of high discharge. In general, 2005 and 2006 were lower flow years than 2000 and 2001, and NH_4^+ concentrations were higher, possibly because NH_4^+ from point sources were not as diluted in 2005 and 2006 as they were in 2000 and 2001.

but were generally lower than estimates from drainage canals or from agricultural fields during fertilization events.

Seasonal variation in N transport in the Yaqui Valley appears to be driven largely by patterns of irrigation and fertilization. We observed a consistent annual pulse of N that coincides with fertilization/irrigation events during the winter wheat season (fig. 10.5). This winter pulse of TDN is largely due to an increase in NO_3^- flux, likely flushed from soils (Harrison and Matson 2003; Riley et al. 2001). This hypothesis is supported by the fact that we have also observed a NO_3^- pulse following summer rains (>20 mm) that are not associated with any fertilizer inputs. We also observed NH_4^+ concentrations to increase following November fertilization-irrigation events in some canals, most likely due to the common practice of wheat fertigation with anhydrous ammonia bubbled into irrigation water. Urea concentrations have also shown a distinct seasonality. The few existing measurements of urea concentrations in drainage canals show an order-of-magnitude increase following valleywide November irrigation and urea addition to fields, suggesting that a significant portion of the seasonal variation in DON is likely due to direct inputs of urea from agricultural fields (Glibert et al. 2006).

In theory, interannual variation in N transport may be affected both by rates of fertilizer application and by the availability and use of water for irrigation. However, this was not borne out in our N transport data. A year with relatively low flow and low rates of N application, such as 2005, demonstrated roughly equivalent rates of N export to other years with higher rates of water and fertilizer use (e.g., 1999–2001). It may be that interannual variation in fertilizer N inputs is dampened by large, and relatively constant, inputs from nonfertilizer N sources such as sewage and manure production. It may also be that the system doesn't respond immediately to decreases in N inputs because the majority of leached N in surface waters comes from soil reservoirs, not from newly applied N. This suggests that in order to reduce N from nonpoint sources, sustained reductions of N inputs or other management approaches (discussed later in this chapter) may be required.

To Coastal Waters

Yaqui Valley drainage waters eventually reach the coast, discharging directly into several coastal estuaries (chap. 7). These estuaries in turn connect with nearshore waters of the Gulf of California, and much of our work in coastal systems of the Yaqui Valley has focused on a large estuary in the south of the Yaqui Valley, Bahía del Tóbari. Highly elevated N concentrations in Bahía del Tóbari suggest that large quantities of N are discharged into the estuary—not surprising given the intensity of fertilizer application in the Yaqui Valley. Based on our monitoring of N concentrations in the valley drainage network, there is every reason to think that many of the Yaqui estuaries receive similarly high N loads, yet a more difficult question to answer is what happens to N entering these estuaries. It is typically thought that much of it should be removed from the estuary via the microbially mediated process of denitrification, which converts NO_3^- ultimately to N_2 gas. This process is often coupled to nitrification (conversion of NH_4^+ to NO_3^-) in sediments, such that NH_4^+ is also removed. Globally, about 40 percent of fixed nitrogen loss occurs in estuarine and continental shelf sediments (Galloway et al. 2004), while within individual estuaries, 10 to 80 percent of N entering these systems is removed via coupled nitrification-denitrification (Seitzinger 1988). Clearly, the uncertainty surrounding the latter estimate indicates that how much N is removed can vary greatly from estuary to estuary; how much N is ultimately denitrified is important for both water quality within an estuary and, potentially, for marine ecosystems downstream.

To gain a better understanding of the process of denitrification in situ, we used molecular techniques to investigate denitrifier communities along a transect in Bahía del Tóbari. Overall, we found extremely diverse communities, which indicates that denitrification is active within the sediments of the estuary (Beman et al. 2008). Denitrification potentials—measured when substrates are supplied in excess—were generally higher at interior sites, where N loading is expected to be greater. Interestingly, the interior sites with higher denitrification potential had less diverse denitrifier communities, which suggested that these communities have adapted in some way to biogeochemical conditions in Bahía del Tóbari.

Based on our limited rate measurements, we estimated—with a number of important caveats[6]—the overall capacity for denitrification within Bahía del Tóbari. Interestingly, we found the percentage of N that can be removed to be relatively low, around 15 to 30 percent of N entering the estuary. This "bottom-up" approach compares remarkably well with another "top-down" estimate based on residence time. As an important related part of the Yaqui Valley project, precise measurements of estuarine bathymetry and circulation were made by physical oceanographers in Bahía del Tóbari (Cruz-Colin et al. in preparation) These data can be used to calculate the residence time for the estuary, which has been shown to correlate strongly with the percent of total N that is removed ($r^2 = 0.88$); longer residence times result in more N being removed (Nixon et al. 1996). Using data from Cruz-Colin et al. (in preparation), residence time within Tóbari is a relatively short five to ten days, which indicates about 14 to 22 percent of N would be denitrified based on the relationship described in Nixon et al. (1996).

Clearly, there is significant uncertainty surrounding both our top-down and bottom-up estimates; however, these uncertainties do not change the overall conclusion: that although a significant percentage of N is denitrified within Tóbari, the majority of N that reaches the estuary is exported. If this holds for the other coastal systems of the valley, 70–85 percent of the N reaching these estuaries flows directly into the Gulf of California.

The Gulf of California is itself extremely nutrient rich, which results in high biological diversity and productivity, and we did not initially suspect that N from the Yaqui Valley would have any appreciable effect on biological processes in nearshore waters. Instead, our aim was to examine eutrophication in the estuaries of the Yaqui Valley, and using satellite imagery, we generated five years (1998–2002) of chlorophyll data—a proxy for algal biomass—to monitor the effects of N from the Yaqui Valley on estuaries. Surprisingly, we found a clear association between the timing of irrigation/fertilization events in the Yaqui Valley and large (54–577 km^2) phytoplankton blooms in the open waters of the Gulf of California, downstream from the estuaries (fig. 10.6). We found that periods of greater irrigation were significantly correlated with these blooms, and when sea surface temperature data were used to account for seasonal variability and the natural process of upwelling, 70 percent of irrigation events in the Yaqui Valley still had blooms associated with them. This research resulted in one of the clearest demonstrations in the biogeochemical literature of the connection between management activities on land and consequences in sensitive ecosystems downstream and was published in *Nature* in 2005 (Beman et al. 2005).

Why does nitrogen from the Yaqui Valley have such a strong effect in the Gulf of California? For the most part, productivity in the ocean is regulated by a balance of three nutrients: N, phosphorus (P), and to a lesser degree, iron. In the Gulf of California, concentrations of N and P are both high and well correlated throughout the water column (Alvarez-Borrego et al. 1978), however there is relatively more P compared to N, such that when all N in surface waters has been consumed by biological activity, there is still an excess of P (Alvarez-Borrego et al. 1978). Even under nutrient-rich conditions, there is an excess of P, suggestive of a persistent N deficit in the gulf (Beman et al. 2005). In effect, N from the Yaqui Valley balances

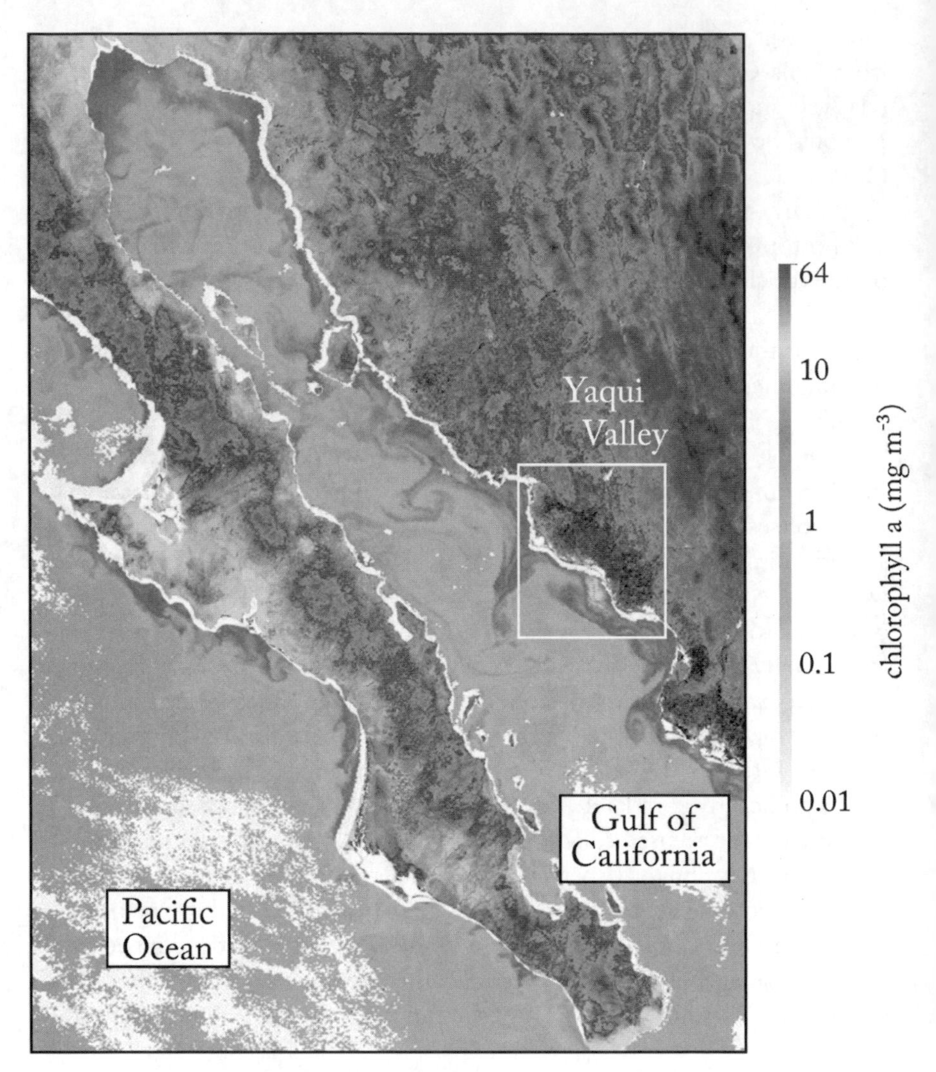

FIGURE 10.6 Satellite imagery taken one day after peak late-season irrigation shows biological activity on land (MODIS Aqua; NDVI) and in the ocean (SeaWiFS; chlorophyll *a*). Phytoplankton blooms, indicated by highly elevated chlorophyll *a* values, are clearly visible in the Gulf of California off the coast of the Yaqui Valley.

this deficit, producing the phytoplankton blooms observed in imagery. Due to the fact that we can place some constraints on how much N might reach the gulf, we calculated that Yaqui N losses during irrigation can support a phytoplankton bloom in the Gulf of California of 50–785 km^2, in close agreement with bloom areas determined using satellite data (Beman et al. 2005). Our overall interpretation of these data is that although N losses from the valley in drainage waters represent a fairly small percentage of N that is applied in fertilizer, very little of this N is removed in coastal estuaries, and the Gulf of California is extremely vulnerable to N in runoff; altogether this results in large blooms of algae in nearshore waters.

Nonagricultural Sources of Nitrogen to Surface Waters

As it became clear that downstream ecosystems were reacting strongly to pulses of N from valley drains, we realized that our view of the Yaqui Valley was dominated by our understanding of nutrient dynamics in agricultural fields. Wheat agriculture is clearly a big driver of N dynamics in the Yaqui Valley, and our studies focused on this sector if for no other reason than what we learned in the fields, drains, and estuaries continually led us to further questions. But how are recent increases in shrimp production, livestock operations, and the human population (fig. 10.7) contributing to N loading in drains and coastal waters? Good estimates of N flows in these sectors still suffer from a lack of comprehensive data, a clear example of how results from basic biogeochemical research in the field leads to changing research priorities, which in turn reveals gaps in our knowledge. In the following sections, we employ the data that we do have with some broad assumptions to make ballpark estimate of the N wastes from livestock production, aquaculture, and human demand. These estimates are summarized in figure 10.3. We have so little data on other potential sources of N in the valley (e.g., industrial waste and vehicular emissions) that even crude estimates were not possible.

Livestock

Livestock production in the Yaqui Valley has increased in the last two decades, most recently due to low feed prices that followed the North American Free Trade Agreement (Naylor et al. 2001). The three most important

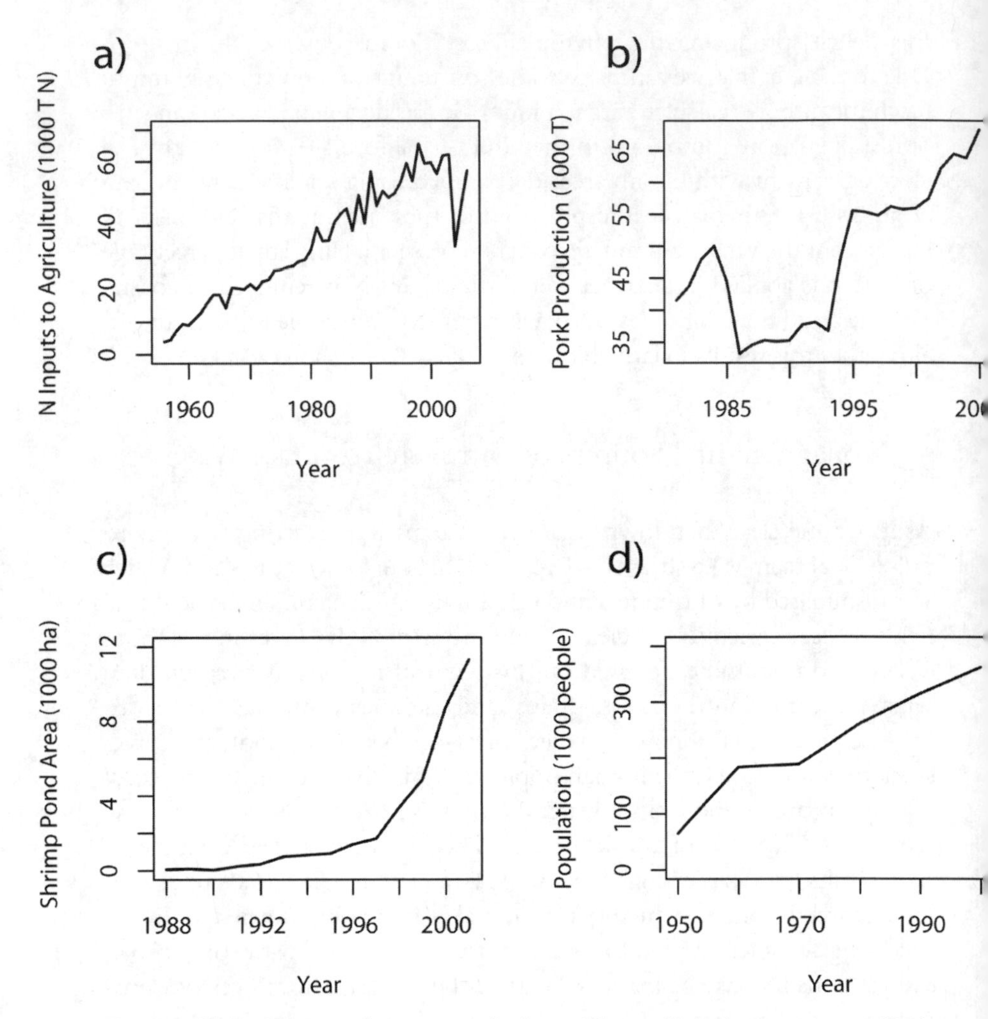

FIGURE 10.7 Nitrogen inputs to agriculture (a) are still the largest source of N in the Yaqui Valley, but recent increases in pork production (b), shrimp pond area (c), and the human population (d), also add to the N burden to the drains, groundwater, and atmosphere.

livestock industries in the valley are hogs, followed by cattle and chickens. The density of hogs in the Yaqui Valley is similar to other hog-producing regions of North America and Europe.[7] The N required to feed the valley's hogs and cattle is less than 10 percent of the fertilizer N applied to valley crops annually (see fig. 10.3). However, approximately 70–75 percent of the N in feed is likely excreted in feces and urine (Smil 2000). If most of this N rich waste flows back into surface waters, livestock operations could constitute an important source of N to surface waters. Livestock waste is typically held in waste-holding ponds for an unspecified amount of time before periodic discharges into valley drains. If all of the N in livestock waste were converted to NO_3^-—unlikely, but the assumption provides an absolute maximum—then the amount of NO_3^- produced by the livestock industries is similar to estimates of NO_3^- leached from all wheat fields in a given year. Livestock are therefore a potentially important source of N in the valley. Open waste lagoons are also a potent source of greenhouse gas emissions (primarily as methane). Interestingly, many livestock producers in the Yaqui Valley are now controlling emissions by covering lagoons and flaring captured methane, funded by foreign companies trying to reduce atmospheric greenhouse gas loads through the Kyoto Protocol's Clean Development Mechanism. The source of feed for the hog industry varies from year to year, but in 2006 hog feed consisted primarily of wheat grown in the Yaqui Valley, and therefore does not represent a new source of N for the valley. This means that the livestock industry is now tightly coupled to the valley's wheat production and associated N requirements. This may change, however, as farmers switch from primarily feed varieties of wheat to varieties that are more marketable for human consumption on the world market.

Aquaculture

Despite a rapid increase in the area of coastal shrimp ponds, shrimp production along the southern Sonoran coast accounted for only 2 percent of N inputs to Yaqui Valley (fig. 10.3). Although no measurements of gaseous emissions of N from shrimp ponds were made in the Yaqui Valley, a study of shrimp ponds in the neighboring state of Sinaloa estimated that a maximum of 39 kg N/ha was lost from shrimp aquaculture ponds through volatilization or denitrification (Páez-Osuna et al. 1997). This represents a small fraction (<1 percent) of N fertilizers applied to agricultural fields in the valley, but if a large portion of this loss occurs as NO_x or N_2O, shrimp ponds might constitute a significant source of these compounds. Shrimp ponds may also play a significant role with respect to coastal N loading. Outlet water from the ponds is estimated to be enriched with as much as a quarter of all DIN delivered to coastal systems from the entire Yaqui Valley drainage network.[8] Although shrimp farm surface area and N input requirements pale in comparison to wheat agriculture, shrimp farms may represent an important source of N to coastal systems due to their proximity to the coast, which leaves little time for processing and removal of exported N before it enters coastal waters.

Human Populations

The Yaqui Valley's human population has increased almost sixfold in the last fifty years, but the N inputs in human diets and outputs in sewage are still small compared to fluxes in the agriculture and livestock sectors. The N contained in the food consumed by Yaqui Valley inhabitants is about a third of that required in feed by the livestock industry (fig. 10.3). Additionally, as of 2005, 70 percent of the valley's approximately 382,000 inhabitants lived in Ciudad Obregón (INEGI 2005), which has been served by two wastewater treatment plants with biological secondary treatment since August 1997. Our calculations suggest that human sewage is not among the most important sources of N on a valleywide scale, although N burdens on drains in communities without wastewater treatment could be locally important.

Yaqui Valley Nitrogen Transformations and Transfers Placed in Context

From the point that N is applied to Yaqui Valley fields, it can undergo multiple transfers and transformations, and integrating across studies in the Yaqui Valley produces an unusually complete picture of the fate of N in the region, from fields to drainage canals to the coast (see fig. 10.3 and table 10.1). In an earlier publication, Ahrens et al. (2008) compared N inputs and N exports from the Yaqui Valley to global model estimates and published studies of N transport from watersheds to the coastal zone. Overall, the percentage of total dissolved N exported from the Yaqui Valley is low; around 4 percent of the approximately 64,000 mg of N inputs (dominated by fertilizer N) is delivered to coastal waters in drainage canals, shrimp pond effluent, and groundwater combined. This low loss rate is likely due to a combination of higher soil N retention resulting from low flushing rates under limited rainfall, and more N processing in streams leading to higher gaseous losses and lower export to receiving water bodies. Subsurface groundwater discharge from the valley to the Gulf of California is also not thought to be a major source of N to coastal waters, contributing less than three percent of surface discharge. On the other hand, limitations in our sampling regime may also have contributed to an underestimation of regional N export. Periodic flushing events (especially the lack of rain event-based sampling) are all probable factors in explaining why estimates of percent N export from the Yaqui Valley are relatively low.

Crosscutting Themes

As with any research effort, we ended more than a decade of research with many questions answered but many new questions unanswered. Gaps in our knowledge base, however, do not override the quite significant set of new knowledge gained about nitrogen processes and transfers from land to ocean and atmosphere. Several key crosscutting themes emerged from our efforts in the fields, drains, and coastal waters. Inputs of N to the Yaqui Valley were dominated by wheat agriculture, and we have shown through field experiments that coupling fertilizer application with crop demand results in decreased N leaching and emissions without affecting crop yield.[9] Fluxes of the greenhouse gas N_2O from both soils and the freshwater system were among the highest ever reported, and N concentrations in both surface waters and estuaries are extremely high (even though the proportion of N inputs to agricultural fields exported to coastal waters is quite low). All of this implies that per-area gaseous and solution losses of N from agricultural fields are quite large. And those losses are not evenly distributed in time and space. For example, discrete valley irrigation and fertilization events result in pulses of gaseous loss, pulses of leaching loss, pulses of flux through drains, and eventually result in pulses of increased productivity in the Gulf of California.

Our earliest interests in the Yaqui Valley were focused on understanding biogeochemical processing in understudied agroecosystems and coupling that knowledge with decisions made on farms. An understanding of such a high degree of spatial and temporal variability was not trivial to capture, requiring time-consuming basic field research. Likewise, our efforts to trace the fate of lost fertilizer, from field to stream to atmosphere to oceans, required the work represented in several PhD dissertations, undergraduate and master's projects, and several postdoctoral projects that ranged over a decade or more. The understanding gained through this research has significantly enriched the field of biogeochemistry. The publication of many peer-review journal articles, including findings published in *Science* and *Nature,* illustrate a reasonably complete picture of N dynamics in agricultural fields, surface waters, and coastal systems. And with our colleagues working in agricultural and economic research, we learned that management clearly influences what happens to nitrogen fertilizer, and that relatively straightforward management changes that increase the precision of application relative to crop demand can significantly reduce losses.

As our knowledge base deepened, we realized that research geared toward policy solutions (rather than basic disciplinary understanding) required acquisition of information that is not traditionally part of disciplinary research. Such information, much of which is still missing in our work, will likely greatly enhance our understanding of links between N cycling and public health concerns in the valley.

For example, the establishment of air quality monitoring networks measuring NO_x, ozone, and fine particulate concentrations in major population centers would be extremely valuable in establishing connections between on-farm activities and public health. Similarly, water quality monitoring of valley wells—including those used for drinking water—is needed to determine the extent of nitrate contamination of the valley's aquifer. Reliable records of public health data from local hospitals would also be valuable. In Bahía del Tóbari and the Gulf of California, little is known about the secondary effects of increased productivity, such as eutrophication, harmful algal blooms (HABs), and oxygen depletion, and how these effects interact with fishing pressures. Lastly, N inputs to the Yaqui Valley are currently dominated by agriculture, but we know less about other sectors (e.g., livestock, aquaculture, and urban) that are growing rapidly and may become increasingly important in the near future. Basic data collection on N management in these sectors is still lacking. As our scale of interest changed, we have realized that more data are needed to address evolving questions geared toward quantifying the true social costs of N inputs in the valley.

The Yaqui N story is an evolving story. It changed, and continues to change, as we expanded our research horizons from fields to landscapes and from land to water and atmosphere. Ultimately, our system-level analyses of biogeochemical transfers and processes that link land and ocean provide a clear argument for the scientific value of place-based, regional studies of ecosystem processes. The nitrogen story also changed in response to the ways in which decision makers in the valley manage their crops, livestock, water, and cities. Images of huge phytoplankton blooms in the coastal waters of the region helped set the stage for a new dialogue on fertilizer use and other issues of sustainability in the Yaqui Valley. Although only a snapshot in space and time, the Yaqui Valley nitrogen story is a parable for global change in the nitrogen cycle. It illustrates the connectivity of the planet—that decisions made in one place and time can influence ecosystems far distant—and that management matters if we wish to both produce food and sustain our land, oceans, and atmosphere.

Acknowledgments

We would like to thank Gustavo Vasques, Ruben González, and Chuy and the gang, for assistance with sample collection and processing; Dolores Vazquez for facilitating the research in Mexico; Tina Billow for getting this work launched; Ivan Ortiz-Monasterio for unwavering help and guidance at every stage; and all other members of the Yaqui team for help and insights throughout the research. We are grateful for funding provided by the Packard Foundation, as well as grants from USDA, NASA, and the Mellon Foundation to PM; NSF graduate student research fellowships to TA and MB; NASA Earth System Science Fellowship to JH; NSF dissertation enhancement awards to MB and JH; and funds from Stanford University.

Chapter 11

Water Resources Management in the Yaqui Valley

GERRIT SCHOUPS, LEE ADDAMS, DAVID S. BATTISTI, ELLEN MCCULLOUGH, AND JOSÉ LUIS MINJARES

Managing water for sustainable use and economic development is a technically and politically difficult challenge for many regions of the world (Lach et al. 2005; Jacobs et al. 2010). Because water is the lifeblood of agriculture in the Yaqui Valley and has been used almost exclusively for irrigation purposes, it is intimately tied to sustainability transitions in agriculture. Moreover, as demands increase for other uses, including supplying urban growth and maintaining environmental flows, broader sustainability concerns are being added.

The story of water resources in the Yaqui Valley is one of evolving institutions, of making decisions amid highly variable climate conditions and of bearing the consequences of management mistakes. Just as responsibility for management of water resources was being handed down from the federal government to a local entity run by farmers, a long string of dry years struck the valley. The drought taught an important lesson about climate variability and management responses and set up a situation in which water managers became receptive to a range of opportunities to improve water management and sustainability.

This chapter provides an overview of water-management issues that Yaqui Valley farmers and managers are facing today. It reviews research efforts that were instigated by these issues, and discusses several promising water-management solutions that could contribute to more sustainable water management in the valley. First, the historical development of irrigated agriculture in the valley is reviewed, including the collapse of agricultural production during the recent drought (1997–2004). Several alternative water-management options are explored in subsequent sections, including better ways of managing surface water reservoirs in the face of future droughts, strategies to increase groundwater use in the valley, and sustainable water management that optimally combines both surface water and groundwater. The chapter ends with a range of recommendations for better water management, and highlights some of the lessons learned for doing interdisciplinary and participatory research in the valley, including suggestions for future work that builds on the results presented here.

The History of Water Resources and Drought

The Yaqui Valley is characterized by a semiarid climate with low average rainfall rates of 300 millimeters, and high evaporation rates of 2,000 millimeters per year (INEGI 1993). Climate is driven by the North American monsoon system, which means that most precipitation falls in the summer months of June through September. Due to these semiarid conditions, agriculture in the valley critically depends on a reliable source of irrigation water. Historically, irrigated agriculture in the valley has relied heavily on runoff from the Yaqui River, which crosses the valley before discharging into the Gulf of California (also known as the Sea of Cortez). The Yaqui River drains an area of approximately 72,000 square kilometers, including portions of the Mexican states of Sonora and Chihuahua, and a small portion of Arizona in the United States (fig. 11.1). In this section, we provide a brief history of irrigation in the valley, discuss the valley's current water resources, and examine the prolonged drought we encountered and addressed during our analysis.

Historical Development of Irrigated Agriculture

Initial efforts at irrigated agriculture were made by the Yaqui Amerindians, who relied on small-scale water diversions to irrigate fields along the river banks (see chap. 2 for more on this period). These agricultural fields are still cultivated to this day and make up the indigenous land reserves (Yaqui Colonies) along the Yaqui River. However, even though Yaqui River runoff is on average sufficient to irrigate the entire valley, it is also highly variable from year to year, thereby greatly reducing the reliability of this source

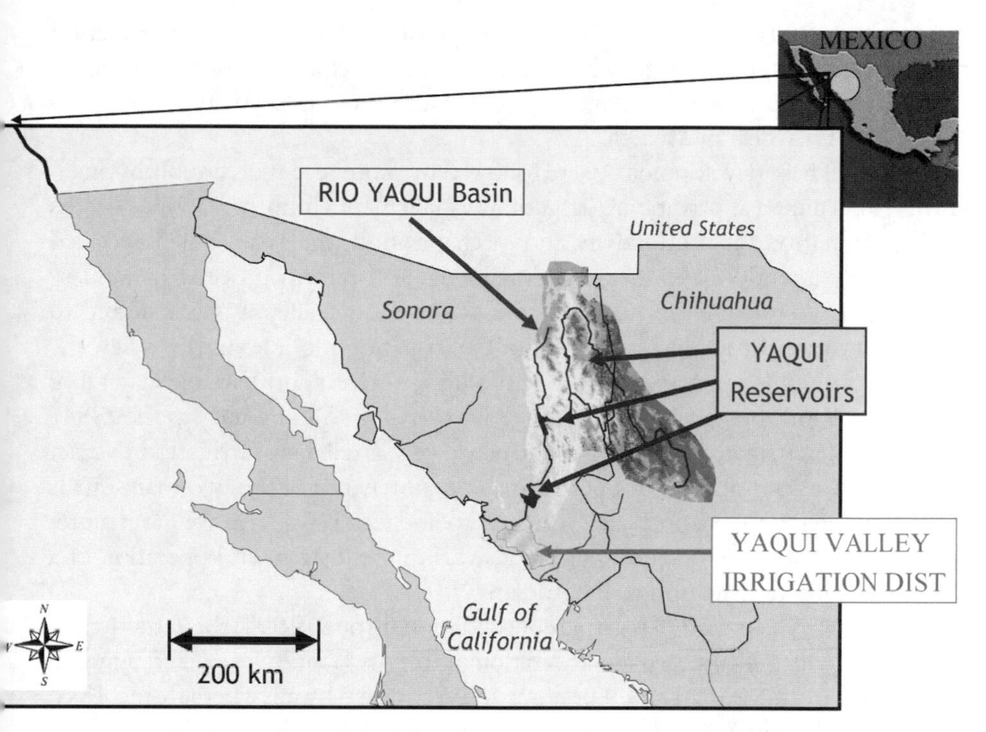

FIGURE 11.1 Map of the Yaqui River basin, northwestern Mexico. The Yaqui River basin encompasses 72,000 km² in the Mexican states of Sonora and Chihuahua (Addams 2004).

of water. Therefore, large scale development of irrigated agriculture, as we know it today in the valley, only became possible after construction of several large dams and water storage reservoirs on the Yaqui River.

The three existing reservoirs, Angostura, Novillo, and Oviachic, were built between 1937 and 1963 and have a combined active storage of approximately 6,000 million cubic meters (MCM), nearly twice the basin's annual mean runoff. Two of the three reservoirs also have hydropower capacity, yet all are managed primarily for agricultural use. The dams were designed by a private company that had constructed the Yaqui Valley's early canal infrastructure around the turn of the century. The Mexican government assumed shares of the corporation by 1936 and began construction of the dams shortly thereafter (Healy et al. 1988). Construction of the first two dams was completed by 1953. These large water storage facilities opened the door to a rapid expansion in total cultivated area, which increased from 50,000 hectares in 1937 to 210,000 hectares in 1955 (Naylor et al. 2001). The abundant availability of a reliable source of irrigation water, combined with wheat research and development of high-yielding wheat cultivars, transformed the valley into the center of the *green revolution* for wheat in Mexico.

These developments were followed by more recent decentralization efforts aimed at privatizing irrigation management (Johnson 1997). With its reputation for productivity and mechanization, the Yaqui Valley received early attention during the national irrigation reform process. In January 1992, the irrigation infrastructure of the Yaqui Valley was transferred to a private entity representing all of the irrigation modules in the valley, the "Limited Responsibility Society of Public Interest and Variable Capital of the Río Yaqui Irrigation District" (known in the Yaqui Valley as the "SRL" or the "Irrigation District"). The newly privatized Yaqui Irrigation District took responsibility for operation and maintenance of the two main canals supplying the Yaqui Valley (Canal Alto and Canal Bajo), as well as maintenance of collector drains and roads within the district, and operation of a number of irrigation wells (Addams 2004).

After a period of internal consolidation through 1994, the Yaqui Irrigation District emerged as a federation of forty-two modules, each ranging in area from 845 to 12,000 hectares and composed by anywhere from 115 to 1,640 farmers (fig. 11.2). Each irrigation module includes all land serviced by a secondary canal network that supplies water from the main canal and delivers it through the network down to individual farms. Each module is autonomous, composed of all water users (farmers) within its particular geographic area who receive water from the secondary canal system. The module covers the operation and maintenance costs of the secondary canal network and holds a proportional water right to reservoir storage each year. Rights do not belong to individual users, but rather to the module as a whole on a proportional basis (Rosegrant and Schleyer 1996; Kloezen et al. 1997).

Even though local management in the valley has been decentralized, operation of the Yaqui River reservoir system remains the sole responsibility of the National Water Commission (CNA). In coordination with local representatives from the irrigation district, the CNA decides how much water is allocated every year for irrigating the valley's crops. These annual allocation negotiations take place in September at the start of the main growing season, and typically revolve around competing risk attitudes and forecasts of future Yaqui River runoff and inflows into the reservoirs. As will be discussed later, reservoir allocations have played a central role during a recent drought (1997–2004).

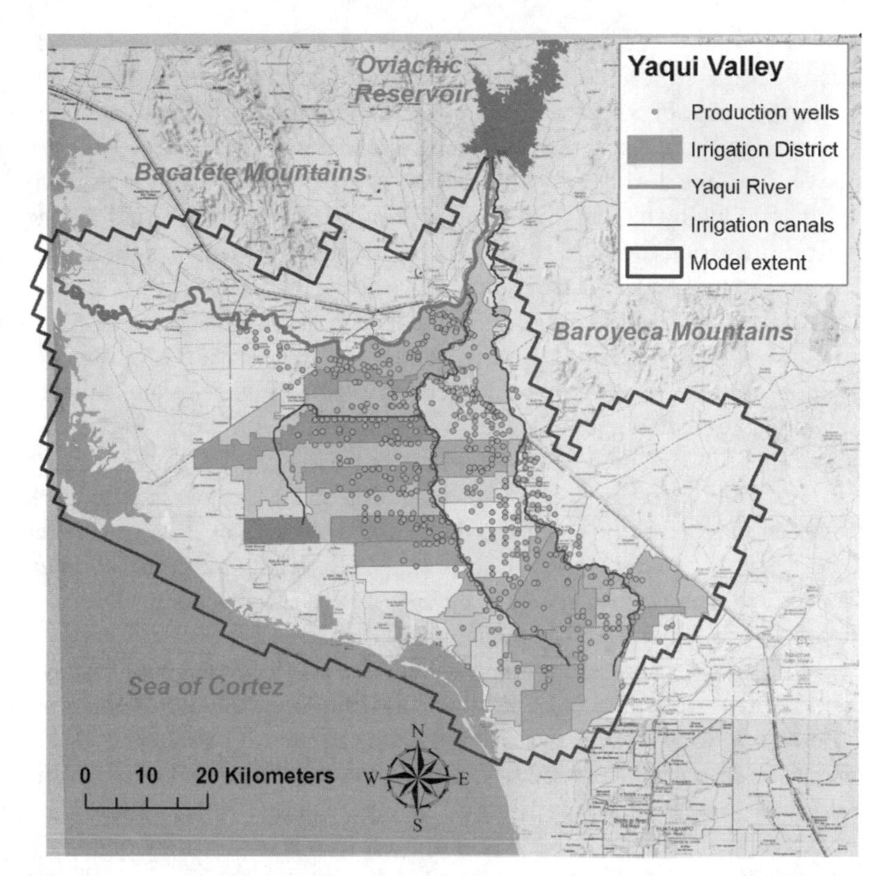

FIGURE 11.2 Extent of the Yaqui Irrigation District along with locations of major irrigation canals and groundwater wells. This figure also shows the extent of the groundwater model that was developed in this project (Addams 2004; Schoups et al. 2005).

The Current Water Resource Situation

In the valley today, irrigable land along the Yaqui River encompasses about 230,000 hectares in the irrigation district (the district), plus an additional 25,000 hectares of Yaqui indigenous reserves. Agriculture is by far the largest water-use sector in the Yaqui River basin, consuming around 95 percent of all allocated water, nearly all of which is dedicated to the Yaqui Valley and nearby indigenous land reserves. Winter wheat dominates cropping patterns in the valley, followed in acreage by a variety of other crops, including maize, safflower, vegetables, alfalfa, cotton, and citrus.

The Yaqui Valley's water resources consist of surface water and the infrastructure to store and transport it, along with groundwater and the wells to pump it (fig. 11.3). The surface water system consists of the three reservoirs on the Yaqui River. Surface water releases from Oviachic reservoir are conveyed to the district by means of two main, open, unlined canals, each about 100 kilometers in length. Along the way, water is diverted by the various irrigation modules that make up the district. Water is further distributed to individual fields within each module by means of a network of secondary irrigation canals, with a total length of 2,500 kilometers. Most irrigation canals are unlined and therefore leak water into the subsurface, leading to water losses of up to 30 percent. Virtually all fields are irrigated through gravity-flow methods (furrows or basins), with the exception of a portion of those planted to fruits and vegetables.

In addition to the network of irrigation canals, a drainage network has been installed throughout the district to drain surplus irrigation water from fields out to the Gulf of California. These drains are primarily open drainage ditches, with a small percentage of subsurface drainage pipes at a depth of 1 to 2 meters below the land surface.

The Yaqui Valley is underlain by an alluvial freshwater aquifer, bordered between the Gulf of California and the Sierra Madre Mountains. Almost 600 wells have been drilled into this aquifer to provide additional water for crop production, although these wells are never all active in the same year. Some of the wells are privately operated, whereas others are managed by the district. Groundwater is pumped into the irrigation canals, where it is blended with surface water for distribution among modules and fields. Historically, groundwater use has been limited due to an abundant and reliable supply of surface water from the Yaqui River basin. A notable exception was the end of the recent drought, when crop production was almost completely sustained by groundwater.

Drought

The 2003–04 growing season brought extreme drought conditions to the valley, rendering reservoirs empty and making surface water irrigation impossible for the season (fig. 11.4). Farmers were able to irrigate less than 30 percent of the valley, and the aggregate value of the Yaqui Valley's agricultural output plummeted to M$383 million, less than 40 percent of the average agricultural revenue during the preceding decade (in real terms).

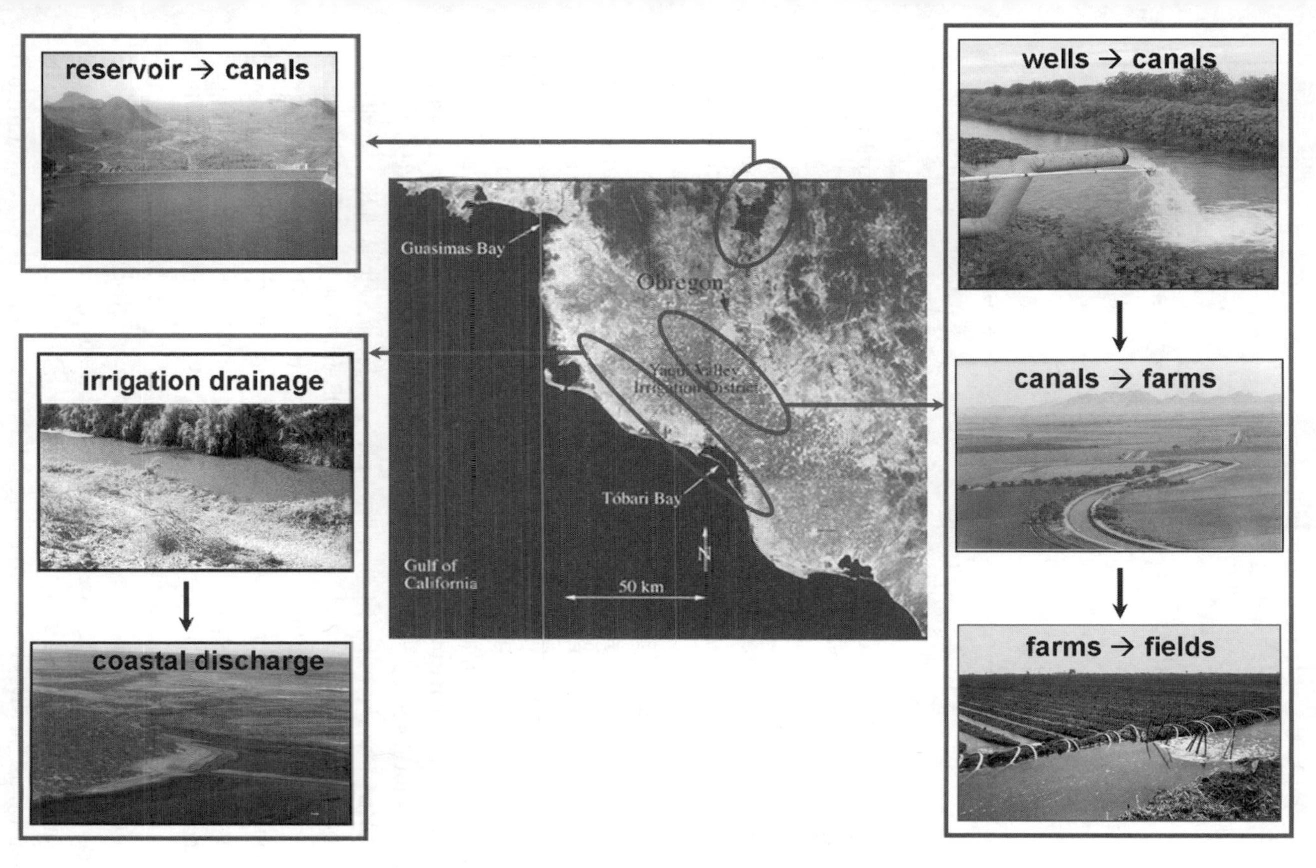

FIGURE 11.3 Water infrastructure and pathways in the Yaqui Valley.

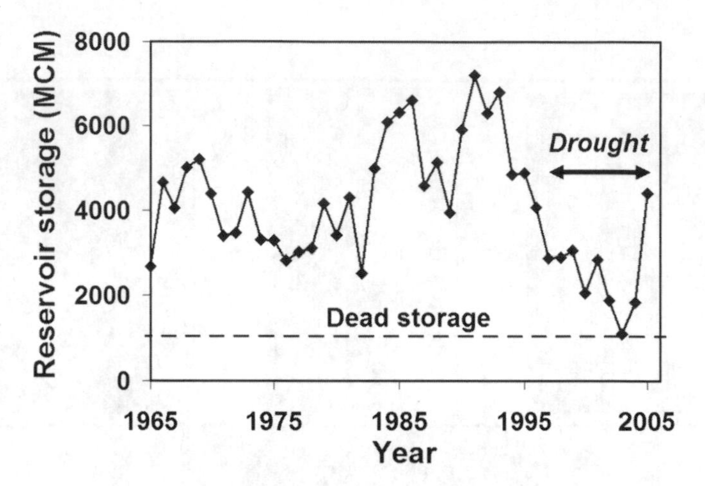

FIGURE 11.4 Annual changes in total water storages in the three Yaqui River reservoirs (MCM is million cubic meters).

Onset of the drought crisis can be directly linked to a shift toward higher-risk reservoir allocation decisions, marked by overly optimistic expectations of future runoff. The increase in risk tolerance coincided with the decentralization reforms described earlier (Addams 2005). During a period of low reservoir levels between 1975 and 1978, the maximum reliance on winter season inflows did not exceed the minimum inflow volume on record. Between 1996 and 2004, the risk component increased, so that there was only a 95 percent chance on average that inflows would exceed allocation. Unfortunately for water managers, between 1997 and 2003, the Yaqui River basin experienced the driest seven-year stretch in winter precipitation since at least 1647 (Diaz et al. 2002; Hulme et al. 1998).

After a string of failed gambles, reservoir levels were at a historical low in the fall of 2002. At the onset of the 2002–03 crop year, the CNA allocated a record low volume to agriculture: 840 MCM, enough to irrigate about 60 percent of the valley. The irrigation district, having requested more water, authorized the sale of planting permits for 200,000 hectares,[1] exceeding "rational" irrigation capacity by 40 percent (Addams 2004). There was an 87 percent chance that in-season runoff would exceed the CNA's allocation, and there was only a 50 percent chance that it would exceed the 101 MCM of expected runoff that the irrigation district had sold to farmers (Addams 2005). According to the National Water Law, the irrigation district was bound to deliver all of the water that it had committed to farmers (CNA 1992). After winter rains failed yet again, the irrigation district installed high-volume pumps to access the reservoir's dead storage capacity to deliver farmers the water they had bought.

With no reservoir storage at the end of the 2002–03 crop year, the valley was in a particularly vulnerable situation for the coming year. After a disappointing monsoon season in the summer of 2003, it became clear that water managers could not repeat their prior allocation decisions. By October of the 2003–04 crop year, the CNA announced that only the Yaqui Amerindians, who by presidential degree have the highest priority of access to water from the Yaqui River system, would receive reservoir water for their territorial lands, located outside of the irrigation district. Based solely on the pumping capacity of the valley's groundwater wells, the irrigation district set out to irrigate the valley. Such widespread use of groundwater and minimal use of surface water was completely without precedent in the valley's history. Historically, surface water has accounted for over 90 percent of irrigation water applied in the Yaqui Valley (Addams 2004).

Need for Research

The impact of the drought highlighted several vulnerabilities that required investigation. It triggered water-management research in three general areas, to be discussed in more detail in the following sections on surface water, groundwater, and sustainable water management.

First, it became clear that management of the surface water system, that is, the allocation of water from the reservoirs, had been too optimistic in that it had relied on projections of future inflows that never materialized. A more sustainable water-management strategy therefore should be based on a rational, scientific assessment of inflow uncertainties and risk-based operating rules that account for this uncertainty. In view of the large uncertainties with regard to reservoir inflows, research was also conducted on possible ways of reducing this uncertainty by providing short-term forecasts of river flows. These topics will be discussed in the section on surface water.

Apart from improving management of the reservoirs, it was recognized that significant potential lies in the augmentation of available water for irrigated agriculture. In general, two strategies can be used to increase water availability and reduce vulnerability to droughts. On the one hand, conveyance and irrigation efficiencies could be increased to improve water productivity and stretch available supplies. For example, irrigation canals in the valley are largely unlined and are therefore expected to leak water into the subsurface, reducing the volume of usable irrigation water. Also, efficiency of current irrigation methods could be improved to further increase water productivity. On the other hand, although the valley is underlain by an extensive alluvial freshwater aquifer, groundwater use has historically been limited due to the availability of abundant surface water from the Yaqui River basin. Significant potential exists for decreasing the valley's vulnerability to fluctuations in surface water supply by relying to a greater extent on groundwater resources.

At the start of this interdisciplinary research project, very little was known about the valley's alluvial aquifer system, including how it would respond to changes in land use, water-use efficiency, groundwater recharge, and groundwater pumping. Therefore, a significant part of the research focused on increasing understanding of the aquifer system, as discussed in the groundwater section.

Finally, building on insights from the previous two sections, the sustainable water-management section summarizes research on several alternative water-management strategies that could make water management in the valley more sustainable and resilient to droughts. Focus is on strategies based on an increase in water-use efficiency and a greater reliance on groundwater for irrigation.

Our research revolved around the development of water-management models to evaluate the potential of both types of strategy. Models were initially developed for the various subsystems, that is, the reservoirs, the canal distribution system, and the groundwater system, and they were eventually linked together dynamically into an integrated water-management model for the Yaqui Valley. The various management models relied invariably upon methods of (1) economic optimization, aimed at identifying water-use decisions that maximize agricultural profits under existing physical and institutional constraints; and (2) physics-based simulation to quantify the effects of these decisions on surface water, groundwater storage, and crop yields. That is, the models were designed to be as realistic as possible by including physical knowledge on hydrologic and agronomic processes, and by accounting for actual decision making of both water managers and water users. Each management model component was grounded in data from the Yaqui Valley, which relied on substantial effort and collaboration with Yaqui Valley water managers.

Surface Water

Median annual Yaqui River runoff amounts to 2,700 MCM. We can compare this number to annual irrigation water demand in the district of about

2,000 MCM and annual water needs of other users in the Yaqui River basin (including urban users, mining, Yaqui Amerindians, and evaporation losses) of 700 MCM. This comparison reveals that in normal years enough surface water is available to support all water users. However, year-to-year variations in runoff, ranging from 1,400 to 6,400 MCM, leave agriculture vulnerable to periodic droughts. Some of these variations have been smoothed out by construction of the three reservoirs on the Yaqui River, with a total active storage of about 6,000 MCM. Based on these numbers, reservoir storage should protect against minor droughts of three to five years, depending on drought intensity. Longer droughts however, such as experienced during 1997–2004, require more careful planning and management in order to prevent drastic declines in agricultural production due to water shortages.

The most important instrument available to water managers in this respect is the annual decision of how much water to release to the district for irrigation. This planning process occurs before the start of the winter wheat growing season in September by a Hydraulic Committee, composed of the local CNA presence and several representatives from the district. The Hydraulic Committee proposes an annual allocation of irrigation water for the district by drafting an irrigation schedule for the entire year based on available water storage in the reservoirs, a forecast of upcoming in-season (October–April) runoff, and an anticipated cropping pattern. The annual allocation proposal is then reviewed and authorized by a federal CNA committee in Mexico City. The proposal also includes an estimated groundwater pumping volume, but historically irrigation has relied heavily on cheap, high-quality surface water, and groundwater has accounted for less than 10 percent of water supply. When available reservoir storage is insufficient to irrigate the entire valley, the stated CNA drought policy is to allocate the available storage plus the minimum in-season runoff volume on record (300 MCM). However, because of negotiation by farmers' representatives serving on the Hydraulic Committee, actual allocations during the drought were typically less conservative than that. This drought management policy has resulted in significant overallocations, causing decreases in reservoir storage (fig. 11.4) and widespread fallowing in the 2003–04 growing season.

In the face of uncertainty regarding future reservoir inflows, a crucial question is how much water to release given initial preseason reservoir storages. One could consider two extreme attitudes. On the one hand, risky behavior results in aggressive allocations and is overly optimistic about future reservoir inflows. The disadvantage of this approach is that it may lead to severe shocks to the agricultural production, as witnessed during the recent drought. On the other hand, a risk-averse policy results in much more conservative allocations, even in periods of abundant water supplies, in order to conserve water for potential droughts. This approach saves water to protect against severe droughts, but likely at the expense of reductions in agricultural production during wet periods.

Minjares (2004) addressed the issue of optimal reservoir operating rules and studied the economic benefits of releasing water from the reservoirs in such a way that agricultural profits are maintained over the long term, especially during dry periods. The study resulted in a series of risk-minimizing reservoir operating rules that have been adopted by the CNA in Mexico. The rules are designed to maintain minimum reservoir outflow levels despite unknown levels of inflow. As preseason reservoir levels become lower, water managers move into lower risk categories, with corresponding reservoir management rules calling for lower reliance on in-season inflows to supplement outflow commitments.

As opposed to adjusting risk attitudes as a function of reservoir storage, another way of improving surface water management is to reduce uncertainties of future reservoir inflows. In other words, if precipitation amounts and resulting future reservoir inflows could be forecasted with a reasonable degree of certainty, then this would benefit reservoir operation by allowing more confident water allocations. Climate research in the Yaqui River basin has suggested that forecasting of in-season winter precipitation and runoff may be possible. It relies on two key observations.

First, even though summertime precipitation is over three and a half times the total wintertime precipitation, winter precipitation is over three times as variable relative to summer precipitation (Nicholas and Battisti 2008). Winter rainfall also correlates more closely with total reservoir inflows than does summer rainfall, probably because rain that falls in the basin during the winter is less likely to evaporate or be taken up by vegetation (fig. 11.5). The Yaqui Valley's most recent drought was in fact associated with deficits in winter precipitation despite average summer rainfall during the period. Climate records show that winter precipitation deficits have played a role in all droughts that occurred over the past century, including the only other drought of comparable severity, which occurred in the 1950s. Reconstruction of a 350-year precipitation record using tree ring data confirms that wintertime droughts recur every 50 years, on average, and that worse droughts than the recent event have occurred during that period.

Second, wintertime precipitation has been linked to the El Niño-Southern Oscillation (ENSO) climate phenomenon, even though summertime

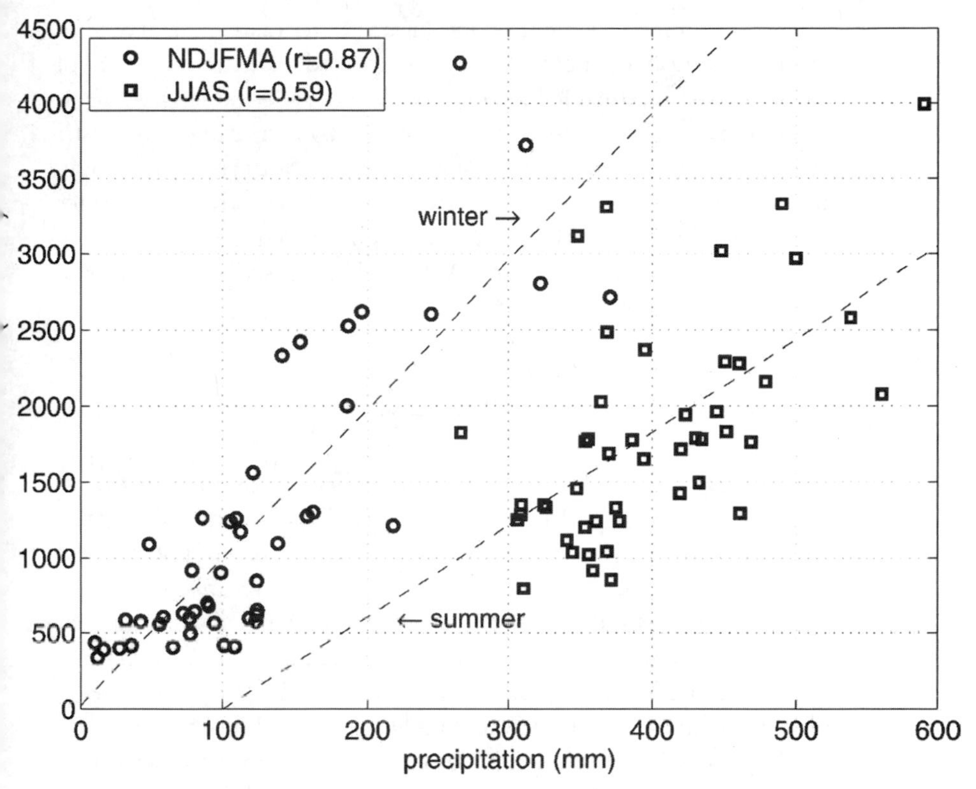

FIGURE 11.5 Seasonal precipitation and reservoir inflow in the Yaqui basin for the period 1956–2002. Dashed lines indicate the best-fit linear relationship between rainfall and inflow. Although the majority of the basin's annual rainfall occurs during the summer months, winter rainfall is more important for the variability in the filling of the Yaqui River's reservoirs (Nicholas and Battisti 2008). © 2011 by the American Meteorology Society.

rainfall appears to be unrelated to any large-scale patterns of variability. The Nino 3.4 index explains over half of the variation in wintertime precipitation (Nicholas and Battisti 2008). Current models allow climatologists to forecast ENSO conditions, projecting the Nino 3.4 index two months in advance with more than 80 percent skill (WCRP 2001).

Hence, given the close relationship between ENSO conditions and winter precipitation, as well as the correlation between winter precipitation and reservoir inflows, there is potential for climate forecasting to play a role in reservoir management decisions in the Yaqui Valley (Nicholas and Battisti 2008).

These developments for improving surface water management are very promising. Additional benefits should be expected by increasing the volume of water that is available for irrigation. There are two major strategies for this: (1) more efficient use of available water by increasing water distribution efficiencies, and (2) groundwater pumping. Both strategies require in-depth understanding of the groundwater system and how it interacts with and depends on the surface water system. Research findings regarding the groundwater system in the Yaqui Valley will be described and summarized in the next section.

Groundwater

Irrigation in the Yaqui Valley has historically relied almost entirely on surface water with minimal groundwater pumping. As a consequence, farmers have been vulnerable to large fluctuations in surface water supply. Research in other irrigated regions has shown that the availability of groundwater may act as a buffer against unreliable and uncertain surface water supplies (Tsur 1990), thereby stabilizing agricultural production. For example, in one region, Bredehoeft and Young (1983) found that groundwater use, in combination with surface water, can double agricultural revenues.

The number of active, functioning pumping wells in the valley during the last thirty-five years has fluctuated between 190 and 500. Many wells have been abandoned due to elevated salinity levels and high pumping costs (González and Marín 2000). Today, groundwater wells are a growing part of the irrigation infrastructure in the valley. Total available storage in the aquifer has been estimated to be approximately 100,000 MCM, which is about sixteen times the available storage in the Yaqui reservoir system (6,000 MCM).[2] Therefore, groundwater could play a central role in a sustainable water-management plan in the Yaqui Valley.

Despite the obvious benefits, potential negative effects of increased groundwater use should be carefully considered. These include depletion of groundwater storage by overdraft, increase of pumping costs due to falling groundwater levels, decrease of crop yields due to irrigation with more saline groundwater (Lefkoff and Gorelick 1990), and deterioration of groundwater quality due to saltwater intrusion (Willis and Finney 1988; Reichard and Johnson 2005). Although most of these issues are of concern in the Yaqui Valley, lack of knowledge about the alluvial aquifer underlying the valley had prevented a quantitative assessment.

A key goal of our research was to quantify how groundwater levels in the Yaqui Valley will be affected by increases in groundwater pumping and decreases in groundwater recharge (e.g., by reducing water losses from irrigation canals). This response depends on two sets of considerations. First, the types of deposits are important, as the response to pumping is different in sandy versus clayey material. Second, the pumping response also depends on how much water enters or leaves the aquifer through other ways. For example, recharge may replenish groundwater storage and compensate for depletion by pumping. Therefore, a groundwater balance that tallies all water flows in and out of the system needed to be constructed.

Information on subsurface deposits was collected from available well logs. Based on this, a conceptual picture was constructed of what the subsurface looks like, as shown by the three different layers in figure 11.6. The first 10 to 20 meters below the surface constitute the shallow aquifer, which receives irrigation-related recharge due to water losses from irrigation canals and from field-level irrigation. During the irrigation season, this recharge causes a seasonal rise in the water table, and subsequent drainage and discharge into the Gulf of California by a network of drainage canals. Water may also be removed by the native vegetation around the canals (phreatophytes) or it can evaporate by capillary rise from the shallow water table. The remaining water either flows laterally toward the coast through the 10- to 20-meter-deep water table aquifer, where it discharges beneath the estuary, or it flows downward. Vertical flow is governed by the underlying discontinuous confining layer (*connector horizon* in fig. 11.6), which ranges between 5 and 80 meters in thickness. The confining layer is highly variable in texture and permeability, consisting of clay in some regions and sand and gravel in others. Finally, the deep aquifer (*aquifer horizon* in fig. 11.6) supplies most of the wells in the valley. Generally, it lies 30 to 100 meters below the surface, ranging in thickness from 30 to 170 meters. The deep aquifer consists of Quaternary alluvium and underlying older unconsolidated Tertiary deposits. Depending on the degree to which the hydraulic permeability of the deep aquifer differs from that of the discontinuous layer above it, the aquifer behaves as either a confined, a semiconfined, or an unconfined system.

The conceptual picture in figure 11.6 formed the basis for a numerical model of three-dimensional groundwater flow (Addams 2004), which was updated and shown to be a quite accurate representation of the real aquifer system by comparing its computations with observed groundwater levels (Schoups et al. 2005). The model includes the spatial variation in aquifer material, and accounts for all major flows in and out of the aquifer system. Therefore, the model is ideally suited to quantitatively assess the effect of changes in recharge and pumping on groundwater levels.

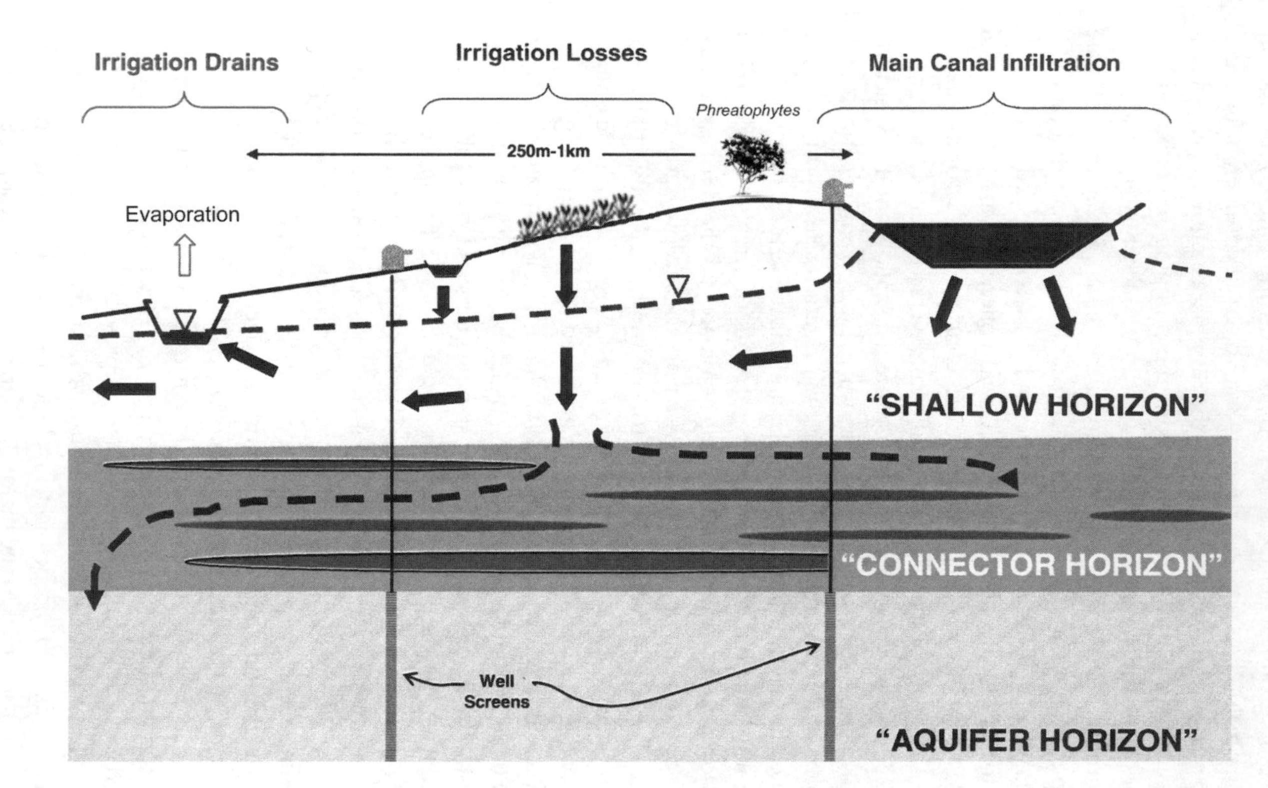

FIGURE 11.6 Conceptual diagram of the Yaqui Valley groundwater system (Addams 2004; Schoups et al. 2005).

Although the basic concept (fig. 11.6) underlying the groundwater model is clearly a simplification of reality, it nonetheless provides the key to understanding groundwater behavior and management in the Yaqui Valley. To a large extent, the shallow and deep aquifers are hydraulically disconnected, in the sense that hydrologic changes in one aquifer have small or negligible effect on groundwater levels in the other aquifer. This insight has two crucial consequences for water management in the valley. First, groundwater pumping from the deep aquifer will mainly affect deeper groundwater levels, with only marginal effects expected on shallow water tables. Conversely, measures aimed at combating excessive pumping lifts and saltwater intrusion will need to directly target the deep aquifer, for example through artificial recharge by deep injection. Second, any reductions in irrigation-related recharge, for example, due to reduction of water losses from irrigation canals or improvements in irrigation efficiency, are expected to mainly affect water tables in the shallow aquifer, without significant effects on groundwater levels in the deep aquifer.

The groundwater model was used to compute a complete water balance of the Yaqui Valley aquifer system (fig. 11.7). The biggest source of water inflow into the groundwater system is *infiltration*, comprising irrigation water applied on agricultural fields and seepage from secondary irrigation canals (84%, on average). The other main inflow source (*canal seepage*) consists of water infiltration losses from the main irrigation canals (16%). Groundwater discharge is composed of crop evapotranspiration (53%), nonagricultural evapotranspiration and bare-soil evaporation (19%), surface and subsurface agricultural drainage discharging into the Gulf of California via the outlets of the drainage network (15%), and groundwater pumping (9%). Subsurface groundwater discharge to the Gulf of California totals 12 MCM/yr, accounting for less than 1 percent of the total annual outflow (Schoups et al. 2005).

It is clear from figure 11.7 that a significant amount of water (19% of total inflows) leaves the aquifer by capillary rise and evaporation from shallow water tables. This process is responsible for soil salinization and is therefore an important management issue. In some areas of the valley, lowering water tables in the shallow aquifer has been necessary to prevent excessive salinization (chaps. 6 and 9). The existing network of drainage canals partially takes care of this, by discharging approximately 15 percent of all groundwater inflows to the Gulf of California. Although drainage is necessary to control shallow water tables, it creates environmental problems in receiving surface waters, as agricultural drainage water contains

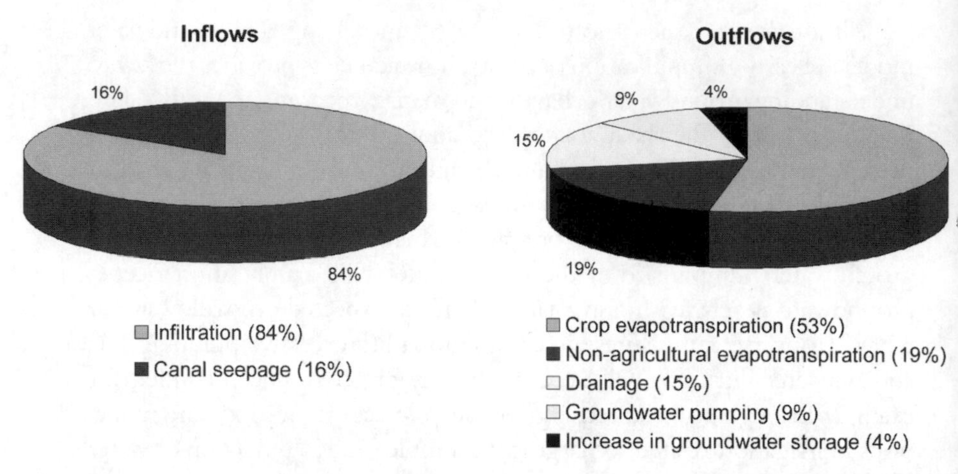

FIGURE 11.7 Groundwater balance of the Yaqui Valley, 1974–1997. Not shown in this figure is groundwater discharge to the Gulf of California, which accounted for less than 1 percent of outflows (Schoups et al. 2005).

excessive concentrations of nitrogen (chaps. 7 and 10). Since irrigation-related recharge is the dominant source of groundwater inflow (fig. 11.7), reduction of recharge by lining of irrigation canals or improving irrigation efficiencies could be an effective measure to control shallow water tables and reduce both evaporation and drainage. This will be addressed in more detail in the next section.

Toward Sustainable Water Management

The aim of this section is to explore and evaluate several alternative water-management strategies that attempt to diminish vulnerability of the irrigated agriculture in the valley to droughts. Research results described in previous sections have highlighted the potential benefits of improvements in surface water and groundwater management. These findings form the basis for formulating sustainable water-management plans. The focus is on an optimal combined use of surface water and groundwater resources, that is, conjunctive use. Considering both sources of water increases flexibility to deal with stress situations (droughts), and therefore greater benefits are to be expected compared to management strategies that look at these two sources separately.

Two sets of management options will be discussed. The first set considers increases in groundwater pumping, whereas the second set looks at ways to improve water-use efficiencies, focusing on water losses from the irrigation canals. In each case, the groundwater model discussed in the previous section is used to quantitatively evaluate potential negative effects of these measures on groundwater levels.

Results in this section are based on quantitative analyses presented by Adams (2005) and Schoups et al. (2006a, b). Their models combine profit-maximizing decision making by farmers and water managers, with physical models that quantitatively compute the impact of cropping and water-use decisions on crop yields, surface and subsurface water levels, and agricultural profits. These integrated-management models were able to mimic historical decision making in the valley, that is, water-use and cropping decisions, as well as resulting impacts on reservoir storages and groundwater levels. As such, they are ideally suited to explore by simulation several alternative water-management options for the Yaqui Valley. The reader is referred to the aforementioned publications for more technical details.

Increased Groundwater Pumping

As discussed in the previous section, there is enough water available in the coastal aquifer to compensate for any variations in surface runoff, as long as enough pumping capacity is available to extract sufficient amounts of groundwater. Here, we evaluate to what extent current groundwater pumping capacity, groundwater salinity, pumping costs, and saltwater intrusion limit groundwater use in the Yaqui Valley.

Effects of Limited Pumping Capacity

The integrated water-management model of Schoups et al. (2006a) was used to evaluate whether the impact of the recent drought could have been avoided if groundwater use had been greater. Simulation results show that groundwater pumping during the period 1996–2005 at a constant rate of 400 MCM, which corresponds to the current full capacity, could have averted the large decrease in agricultural production during the 2003–04 season (fig. 11.8). Increased groundwater use, even during wet periods, saves more surface water for future droughts, and hence decreases vulnerability of agricultural production. The model simulations showed that

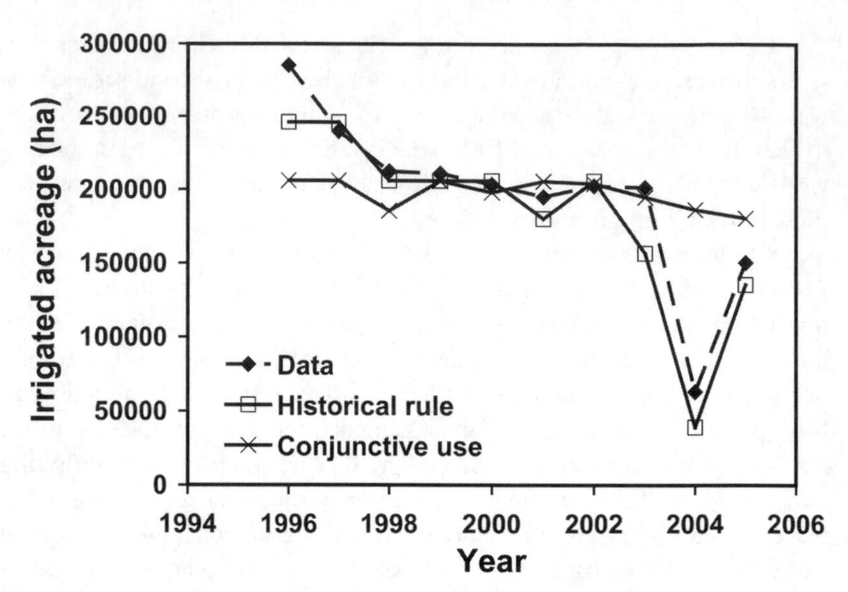

FIGURE 11.8 Effects of increased groundwater use on irrigated acreage in the district. The "historical rule" corresponds to historical water management with low groundwater use, whereas "conjunctive use" simulates groundwater use at a maximum rate of about 400 MCM/yr. The latter strategy maintains agricultural production throughout the entire drought (1997–2004) (Schoups et al. 2006a).

groundwater use at the current capacity of 400 MCM/yr has no serious adverse effects on crop yields, pumping costs, or farm profits, and does not cause saltwater intrusion.

This is an encouraging result as it shows that a balanced use of surface water and groundwater resources may result in more sustainable management. However, since the current capacity of 400 MCM/yr is not sufficient to irrigate the entire valley, groundwater pumping needs to occur during both wet and dry periods, as this is the only way to save enough surface water to complement groundwater pumping and satisfy irrigation water demand during extended droughts. This may cause problems during wet periods: since groundwater pumping leaves more surface water in the reservoirs, it inevitably leads to wasteful reservoir spills during periods of excess surface runoff. This trade-off has been described in detail by Schoups et al. (2006b). An option would be to convey the excess water to the valley for artificial recharge of the deep aquifer-using injection wells. The added benefit is that groundwater levels could be restored in this manner.

Effects of Pumping Costs and Saltwater Intrusion

An alternative option is to install additional wells in order to increase groundwater pumping capacity. Groundwater would then be used only during droughts, whereas surface water would be used as long as it is available. In theory, if sufficient pumping capacity is available, then any drought could be accommodated. The question is whether groundwater salinity, pumping costs, and saltwater intrusion would then become limiting factors. In order to evaluate the potential for groundwater as the main irrigation source in the Yaqui Valley, the integrated water-management model of Schoups et al. (2006a) was run for different "additional-well" scenarios, corresponding to additional installed pumping capacities of, respectively, 50, 100, 200, and 300 MCM/month.

Figure 11.9 shows how simulated irrigated acreage changes as a function of added pumping well capacity. Results are shown for cases with and without accounting for pumping cost, and with and without considering saltwater intrusion. When pumping cost and saltwater intrusion are not considered, then irrigated agriculture in the valley can rely entirely on groundwater, as long as enough additional wells are installed. This occurs for an additional capacity of 300 MCM or 1,200 wells.

If we change the model to account for both saltwater intrusion and higher energy costs due to pumping, then the picture changes completely. During the 2003–04 growing season, the risk of saltwater intrusion severely limits agricultural production (fig. 11.9). In this case, the model automatically restricts groundwater pumping in irrigation modules that are located near the coast in order to prevent saltwater intrusion. During the 2004–05 growing season on the other hand, the model indicates that pumping costs become too high to profitably grow crops over the entire district (fig. 11.9). These results indicate that it is unlikely that irrigated agriculture in the valley can be sustained on groundwater as the only source of irrigation water. First of all, the valley's proximity to the Gulf of California puts a limit on how much groundwater pumping can occur without causing saltwater intrusion. Secondly, historically, wheat profit margins have been too small for farmers to be able to afford higher pumping costs caused by large groundwater pumping lifts.

It may however be possible to circumvent some of these constraints. For example, a rotating pumping schedule could be implemented to minimize saltwater intrusion, but such a strategy would require careful monitoring and coordination at the district level. Greater crop profit margins may be possible by a gradual shift in crop production from traditional

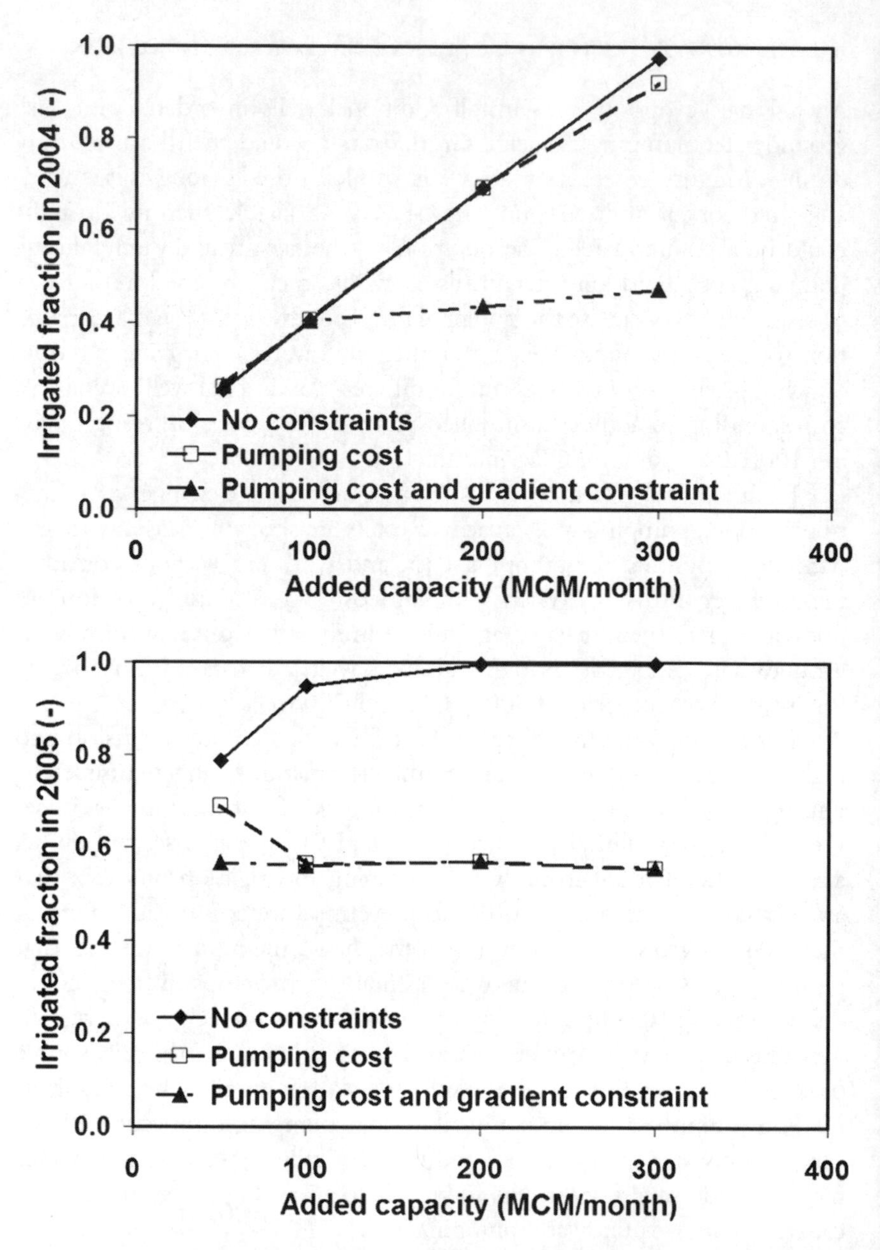

FIGURE 11.9 Effects of additional installed pumping capacity on (a) irrigated acreage in 2004, and (b) irrigated acreage in 2005. Simulation results are shown for cases with and without accounting for pumping cost, and with and without considering the seawater intrusion gradient constraint. Irrigated acreage is expressed as a fraction of the total irrigable land (Schoups et al. 2006a).

winter wheat cultivation toward higher-value crops, such as vegetables and fruit (Addams 2004). However, these crops are much more sensitive to salinity, and therefore irrigation with groundwater may become a problem, as discussed next.

Effects of Groundwater Salinity

Groundwater is more saline (1–4 dS/m) compared to surface water (0.5 dS/m). Elevated salinity may negatively impact agricultural production (crop yield), although this depends to a large extent on how sensitive the crop is to salinity (Maas and Grattan 1999). Data on specific crop salt tolerances were used to evaluate potential yield decreases for Yaqui Valley crops (Addams 2004). It was found that winter wheat varieties cultivated in the Yaqui Valley are quite salt tolerant and are not expected to suffer any negative impacts when irrigated with typical groundwater salinities of 1–4 dS/m. However, other crops, such as maize and vegetables, are much more salt sensitive. Therefore, irrigation of these crops using undiluted groundwater may lower crop yields, or worse, lead to crop failure.

There are various management options to deal with this. First, many groundwater wells pump water directly into the irrigation canals. This means that more saline groundwater can be mixed and diluted with fresh surface water from the reservoirs before being applied to agricultural fields. A disadvantage is that fresh water becomes more saline. Second, greater irrigation water salinity can be partially compensated for by increasing irrigation amounts. In other words, a crop irrigated with groundwater may achieve the same yield as one irrigated with surface water, as long as groundwater irrigation amounts are sufficiently increased to compensate for its higher salinity. Therefore, we explored methods to equitably distribute irrigation water given that some farmers are more likely to receive water of increased salinity due to their location in the valley and along the canals. Model simulations by Addams (2004) showed the feasibility of such an equitable policy, which distributes surface and groundwater among modules and farmers based on an equal-yield rather than an equal-volume basis. With such a system, users who receive a greater proportion of saline groundwater are compensated with a greater total allocation and suffer no yield declines in comparison to other users.

However, greater reliance on more saline groundwater may exacerbate soil salinization in areas with inadequate drainage. As noted in chapters 6 and 9, clay soils in low-lying areas may have poor drainage and high water tables and are therefore vulnerable to salinization. Although our models did not evaluate the longer-term consequences of saline groundwater application to these soils, it is clear that the absence of adequate drainage presents a risk for these regions.

Increased Water-Use Efficiency

The second set of management options looks at ways of improving water-use efficiencies within the district. In principle, this can be achieved by (1) allocating water to those areas where it is expected to generate the largest economic profit, and (2) by minimizing water losses during distribution of water from the reservoirs into irrigation canals and onto agricultural fields.

A common strategy to improve water-use efficiency and promote redistribution of water to where it generates greatest economic benefit is to implement a water market. This allows farmers to sell some of their unused irrigation water to others who experience a deficit. Current district policy allows such water transfers between farmers and between modules, as it can offset water deficits in areas with limited groundwater pumping capacity. Addams (2004) investigated the benefits of this strategy using a management model designed to mimic such market-based water transfers. The amount of transferred water predicted by the model compared favorably with that recorded by the district for the four-year period 1997–2000. However, the economic benefit of the water market was small, generating an overall profit increase of just 1 percent.

Another option for improving water-use efficiency within the district is to reduce water losses by improving conveyance efficiencies. In the Yaqui Valley, the irrigation canals that distribute water from the reservoir to individual farms are earthen, unlined canals that leak water to the aquifer. Consequently, the water distribution system in the Yaqui Valley is quite inefficient and lining the irrigation canals may reduce infiltration losses and save more water for irrigation. Overall conveyance efficiency of the main irrigation canals is about 81 percent, whereas conveyance efficiency of secondary canal network varies spatially throughout the district (fig. 11.10) from 70 percent to 90 percent. An obvious remedy to avoid these infiltration losses is to line the canals. Further benefits may come from improving field irrigation methods, as current efficiencies are also fairly low (fig. 11.10; also see chap. 9).

Simulated results with the integrated-management model of Schoups et al. (2006a) show that lining the irrigation canals has a significant and positive impact on agricultural production during droughts. Lining the

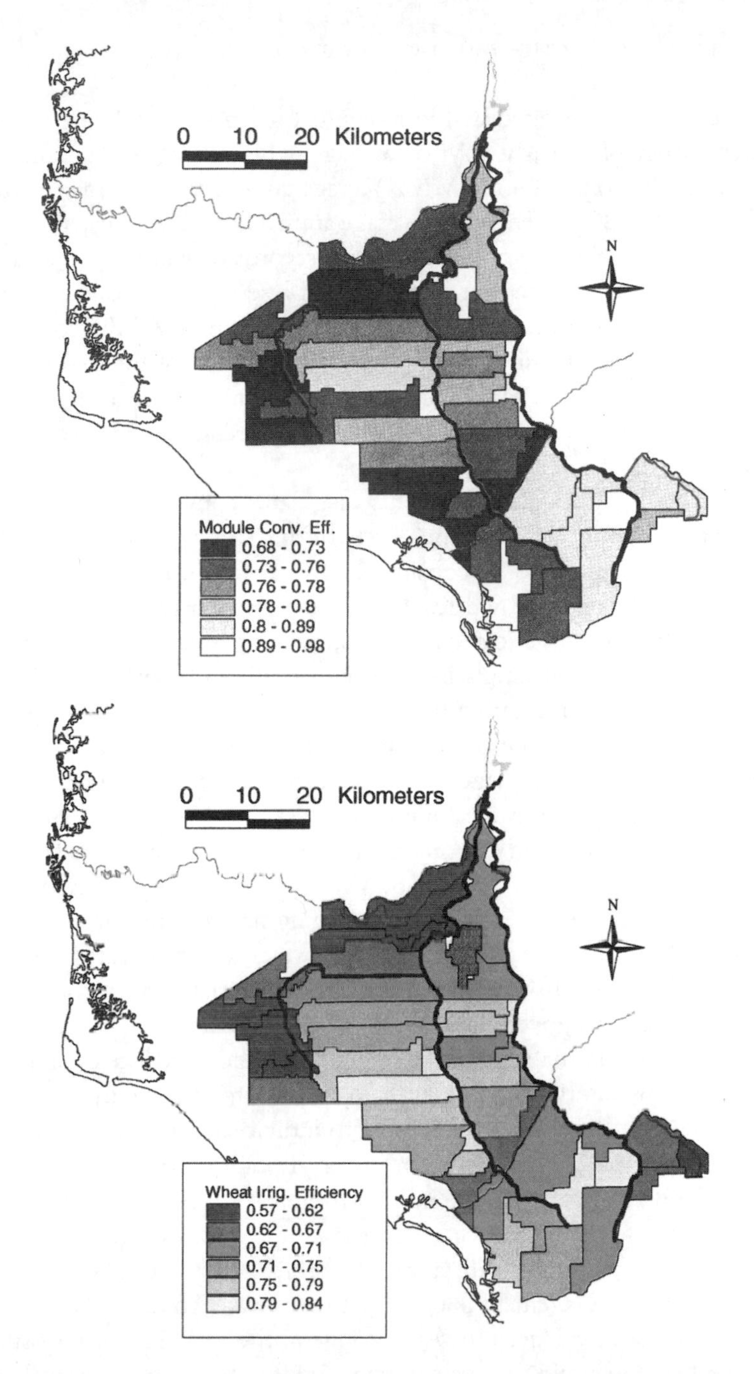

FIGURE 11.10 Spatial maps of conveyance efficiency of secondary irrigation canals, and wheat irrigation efficiency, for each irrigation module in the district for the period 1996–2001 (Addams 2004).

secondary canals, located within individual irrigation modules, results in water savings of 294 MCM/yr on average, whereas lining the main canals yields an additional 288 MCM/yr. Average annual water savings from lining amount to 30 percent of the average annual historical reservoir releases. This number corresponds well to district records that indicate that during that same period 33 percent of the annual reservoir releases were lost by infiltration losses from the secondary and main canals. All this extra water becomes available for irrigation, either for direct use or for storage in the reservoir and later use. Model simulations indicate that the drop in agricultural production during 2003–04 could have been avoided if conveyance efficiencies had been improved by lining.

However, although the benefits of greater conveyance efficiency for agricultural profits are clear, the effects of this strategy on the groundwater system also need to be considered. In particular, reduced infiltration losses from the canals could result in reduced recharge to the underlying aquifer, which may lead to declining groundwater heads and greater pumping costs. Our model simulations show that although canal lining reduces groundwater recharge by ~50 percent, this has no significant effects on groundwater levels in the deep aquifer. Instead, reduced recharge results in reduced agricultural drainage and evaporation from the shallow water table and also lowers groundwater pumping rates. In other words, canal lining appears to be a win-win option: it makes more water available for irrigation, it decreases shallow water table evaporation and therefore soil salinization, it reduces environmentally damaging agricultural drainage to the Gulf of California, and it reduces groundwater pumping rates and costs. An important drawback is that it requires an initial investment that will require substantial government support.

A cheaper alternative to canal lining would be to reduce canal infiltration losses by altering surface-water distribution patterns in the district. As it is now, surface water is allocated to individual modules proportional to their irrigable areas. However, water allocations to modules located further downstream along the irrigation canals are characterized by larger losses and lower conveyance efficiency, simply because water has to be transported over a greater distance. Therefore, losses could be reduced by a clever water-management plan that allocates water to those modules that can be served most efficiently. Any change in the spatial pattern of surface water allocation could then be compensated for by groundwater pumping. Addams (2004) investigated this option using a water distribution model for the district. Results showed that overall conveyance efficiency to the modules could be raised from 81 percent to 88 percent by optimally *flexing* the location and timing of groundwater pumping to minimize infiltration losses from the irrigation canals. Such a strategy essentially comes down to greater groundwater use in modules located downstream along the main irrigation canals. A disadvantage would be that groundwater pumping becomes concentrated in a smaller area of the district, thereby locally amplifying any negative effects of groundwater pumping, such as greater irrigation water salinity and greater pumping lifts.

Conclusions and Recommendations

Large-scale irrigated agriculture in the Yaqui Valley would not have been possible without a reliable supply of freshwater provided by construction of reservoir storage capacity on the Yaqui River. Together with wheat-breeding programs and an intensive use of fertilizer, the availability of freshwater turned the valley into the center of the green revolution for wheat, with the valley producing some of the highest wheat yields in the world. Despite these advances, the vulnerability of current water-management practices has been exposed during the recent drought, which resulted in widespread land fallowing due to water shortages. It also highlighted the need for careful management to prevent soil and groundwater salinization in this coastal valley.

The drought crisis stimulated research aimed at understanding the interplay between hydrologic, agronomic, and economic aspects of irrigated agriculture. This has resulted in several key insights and promising solutions toward more sustainable water management in the valley. The purpose of these concluding remarks is to summarize significant research results discussed in this chapter, highlight lessons learned from doing interdisciplinary research in the Yaqui Valley, and suggest promising directions for future work.

Summary of Research Outcomes

Impacts of a recent drought (1997–2004) were caused by a combination of risky surface water management, coinciding with decentralization of water management in Mexico, insufficient groundwater pumping capacity, and an inefficient water distribution system.

Future drought shocks could be alleviated by better risk management of surface water resources. Such risk-based reservoir operating rules, which determine how much water to release from the reservoirs based on preseason reservoir storage, can mitigate some of the impacts of major droughts. This may be further helped by short-term forecasting of winter precipitation and reservoir inflows by exploiting a connection between predictable larger-scale climatic variations and winter precipitation in the Yaqui River basin.

Our research results suggest that further significant improvements in sustainable water management in the Yaqui Valley are possible by two strategies, namely, (1) relying more heavily on groundwater, and (2) improving efficiency with which water is transported to fields. Both strategies could significantly affect the groundwater system and insights gained by hydrologic research in the valley have elucidated what those impacts would be. These insights have shown that there is a disconnection between the shallow and deep aquifers, and that evaporation from shallow water tables is a key process in the overall groundwater balance of the valley. Since evaporation causes soil salinization, measures should be initiated that control and lower shallow water tables. Given the disconnection between shallow and deep aquifers, this is most efficiently done by reducing groundwater recharge, for example, by minimizing water losses from canals and fields into the subsurface.

We found that greater reliance on groundwater may smooth out fluctuations in surface water. With current limited pumping capacities, this can be done without any significant negative effects. However, when groundwater pumping capacity is increased, then both pumping costs and saltwater intrusion become important limiting factors. The increased energy costs factor into on-farm profitability and crop decision making, something to be taken seriously, as major changes in the price and subsidy environment have occurred since the last period of significant groundwater extraction in the 1970s. Declines in groundwater levels by pumping could potentially be alleviated by injecting excess freshwater into the deep aquifer during wet periods.

In the majority of land in the valley, increases in irrigation water salinity due to increased groundwater use are likely not to be a limiting factor on crop production, at least not as long as wheat is the dominant crop. If there were to be a switch to more salt-sensitive crops, careful salt management will become more important. This could be achieved by mixing fresh and saline waters or by implementing a water distribution plan based on equal crop yields instead of equal water allocations.

Our research also highlighted the broad scope for improvements in conveyance and field irrigation efficiencies. A retrospective analysis of groundwater-canal operations by the irrigation district found that a 7 percent improvement in water efficiency was possible by optimally managing well pumping and surface-water deliveries to irrigators. A more dramatic impact can be expected by increasing water-use efficiencies through lining of irrigation canals. This is likely to have the greatest positive impact and can be considered a win-win strategy: (1) it reduces water losses, so that more water is available during droughts, (2) it reduces shallow water-table elevations, resulting in less evaporation and soil salinization, (3) it reduces agricultural drainage, and possibly also nitrate loading to the Gulf of California, and (4) it does not lower groundwater levels in the deep aquifer, instead reducing the need for groundwater pumping.

In summary, the solution toward more sustainable water management in the Yaqui Valley will likely consist of a combination of several strategies. On the one hand, infrastructural changes, such as lining of irrigation canals and installation of additional groundwater wells, promise to provide considerable benefits. These changes should however be accompanied by careful monitoring and knowledge-intensive management approaches that consider the spatial and temporal variation of water resources in the valley, the Yaqui River basin, and the larger-scale climate system it depends on.

Lessons Learned

Beyond assessing the systems and identifying sustainable strategies, a principal goal was to position our research to effect changes in water management. To that end, we did our best to engage local water managers in our research, with mixed results. Other researchers and practitioners have found that involvement of local water managers in research helps set a better course toward the future (Lebel et al. 2005; Richter et al. 2006; Tidwell et al. 2004). We sought an active dialogue with local partners for several reasons. The large amount of historical data required for policy analysis, such as hydrologic observations, historical agricultural prices and input costs, and operational data, required a two-way relationship with local managers. Further, we hoped that active involvement with local partners would help define the scope of solutions that would be actionable as well as optimal in an engineering sense. This would help to avoid solutions and recommendations that are too complex to implement in the political realm.

We built relationships with local academics and water managers in the irrigation district and were successful in interacting with several stakeholders at the operational level. However, earlier and closer engagement of the irrigation district could have generated stronger partnerships with the management personnel of the district.

Nevertheless, several research results have been adopted or transferred to water managers in the valley. First, surface-water allocation rules emerging from the modeling effort have been adopted by Mexico's CNA and the Yaqui Valley's irrigation managers. Efforts continue to incorporate reservoir inflow forecasts into the model and to make more of the model's components accessible to the valley's decision makers. Further, a user-friendly interface and manual for the integrated water-management model has been developed and made available to water managers and farmers in the Yaqui Valley. This water-management model integrates our insights obtained over the years and can be used by water managers to explore alternative water-management strategies.

It is encouraging that infrastructural changes such as lining of irrigation canals and additional groundwater wells were already being planned in a 2003 district modernization plan (Distrito de Riego del Río Yaqui 2003). Combined with knowledge-intensive management tools generated by our project, this should put the Yaqui Irrigation District on the right path toward more sustainable water management.

Future Directions

Although work reported in this chapter was based on an integration of hydrologic, agronomic, and economic aspects of irrigated agriculture, tremendous opportunities exist for further integration of disciplines, based on insights discussed in other chapters of this book.

- Attention has largely focused on district-level improvements in water management, without thoroughly studying potential improvements in field-level irrigation efficiencies. These topics are partly addressed in chapter 9. Alternative field irrigation strategies that use less water, either by using fewer irrigations or by investing in sprinkler and drip systems, could be integrated into the management model presented in this chapter.
- An assessment of the long-term impacts of increased groundwater use on soil salinization, especially in low-lying areas with inadequate drainage, is needed. This will provide a more complete picture of the feasibility of groundwater irrigation in the valley.
- Our results suggest that a greater dependence on groundwater may

require a shift from wheat-based agriculture toward higher-value crops. Also, cultivation of less water-intensive crops may be worth considering. Integration with economic, social, and policy studies is necessary to assess whether such shifts are possible.

- Coupling of the integrated water-management model with climate forecasting should be done. Forecasting can reduce reservoir inflow uncertainties and is therefore expected to result in more efficient reservoir operating rules.
- Opportunities exist for integration of fertilizer management and nitrogen cycling in the valley (chap. 9) with water and salinity issues addressed in this chapter. To this end, nutrient-cycling models developed as part of the Yaqui research project could be incorporated into an integrated decision and management model. This would further broaden the picture of agricultural and environmental sustainability of the Yaqui Valley.

Acknowledgments

We thank Steven Gorelick and Ivan Ortiz-Monasterio for their assistance in many aspects of this analysis, and Walter Falcon and Pamela Matson for ongoing input to the research and comments on this draft. We thank the Yaqui Irrigation District for sharing irrigation records and other useful data. In addition, we are grateful to Pedro Valenzuela Zárate and Ruben Rodrigo Garcia for taking the time to meet with us and for answering many of our questions regarding irrigation and water management in the valley. Ing. Ramón Romero Arreola, president of the Módulo de Riego No. 12, and the personnel of the CNA Departamento de Hidrometría del Distrito de Riego No. 041, Río Yaqui are also kindly acknowledged.

PART IV

Conclusions

Chapter 12

Lessons Learned

PAMELA MATSON, ROSAMOND NAYLOR, AND IVAN ORTIZ-MONASTERIO

As it evolved over fifteen years, our research effort in the Yaqui Valley grappled with a range of issues related to land and water management, from the policy environment to constraints on resources to consequences for local people and ecosystems. The preceding chapters of this volume illustrate crucial interactions within the human-environment systems of this valley, across agricultural, water, coastal and urban, and natural ecosystems, and across space and time scales. The research suggests that management issues must be studied with these connections in mind, and that systems perspectives are key to developing solutions that make sense for both people and the environment.

In the process of our research, and in attempting to link our research with decision making, we learned a number of lessons. Some of these lessons are apparent in the earlier chapters; we use this final chapter as an opportunity to highlight the most important ones. We believe that many of our conclusions can be generalized beyond the Yaqui Valley to other regions where resource sustainability and economic sustainability are the emerging bywords of the day, and to other research groups that are trying to mobilize a sustainability transition. The next two sections of this chapter address lessons about improving understanding and translating knowledge into action.

For academic researchers, our reflections on research approaches, training students, and institutional organization may be of interest. In the latter half of this chapter, we discuss the experience of doing this work and its impact on sponsoring institutions. Other experiments like this one are playing out around the world, and we hope our ideas contribute to a growing understanding of what is needed to engage academic researchers in sustainability science—the broad field that focuses on interactions within human-environment systems, and on the knowledge, tools, technologies, and approaches needed to meet sustainability goals (Kates et al. 2001).

Lessons for Sustainability Transitions in the Yaqui Valley and Beyond

Sustainability is an ongoing process; research and management capacities are needed to continuously advance sustainability goals.

Seeds of a sustainability transition in the Yaqui Valley can be seen in efforts to (1) increase agronomic yields and profitability through modern agricultural technology and knowledge-intensive management approaches, (2) increase nitrogen use efficiencies and reduce nitrogen and other pollutant losses, (3) increase the sustainability of water resources through management approaches across a range of scales, and (4) identify and reduce vulnerabilities to changes, including climate changes. However, much more remains to be done in these areas, and in some ways our work has just skimmed the surface. Efforts in other sustainability pathways—such as diversification to different, more site-optimum crops; landscape-scale management that maximizes ecosystem services by matching crop and soil types and by embedding conservation strategies in the agricultural landscape; the development of sustainable livestock management approaches; or the development of alternatives to agriculture altogether—have received very little attention but could be part of the sustainability transition in this place. One could argue, in fact, that our research effort, which focused on the dominant current practices and led to evolutionary, incremental, near-term improvements, should instead have been focused on the longer-term, more significant, more revolutionary and game-changing alternatives. This was a trade-off that we contemplated and discussed, but we never seriously considered shifting our focus. As researchers and educators who wished to make a difference, we traded off the longer-term and potentially more optimal, but much less certain, solutions for the ones that we thought we could deliver, even knowing they were only first steps. At the same time, our near-term efforts changed the vocabulary and the agenda of the valley to one that included the idea of sustainability; that change may be the most critical foundation for the larger, longer-term efforts that need to come.

One of the benefits and also one of the drawbacks of working in one place for an extended period is recognizing all that *could* be done, if there was only enough time, people, expertise, and money. Perhaps that is one of the lessons of a sustainability transition: it is never done. Based on our experience, it is about moving one step at a time, hopefully more forward than backward, learning as we go. The research needed is not a one-shot deal. This perspective speaks strongly for building capacity for research in a place, and luckily, the Yaqui Valley has a strong local base of scientists and managers who are engaged in the transition.

One of the most critical barriers to sustainability transitions in the Yaqui Valley is the same challenge that faces all of agricultural decision makers: the need for decision making under uncertainty.

Farmers in the valley face heterogeneity, variability, volatility, and uncertainty in soil, pests, insects and diseases, climate conditions, water resources, national policies (applied to the country as a whole, despite vast differences between the subsistence-scale and high-productivity systems in Mexico's dual agricultural sector), international policies, international markets and contracts, commodity and input prices, access to credit and inputs, environmental awareness, social contexts, and information. Unlike most other sectors of the economy, agriculture has a unique temporal dimension; farmers make most of their key decisions before or at the time of planting, prior to the emergence of biotic and abiotic stresses and often months before the crop goes to market. Part of the problem is incomplete knowledge, and part of it is simply that some of these things are inherently variable. Our work suggests that adding new goals—and new ways of doing things—to this complicated and uncertain world can be very difficult and even risky.

On the other hand, to the extent that research can focus on new approaches that help decisions makers deal with uncertainty, sustainability goals can benefit along with others. For example, our work on water and nitrogen-use efficiency approaches that use real-time information represent *no-regrets* actions that can be taken at little cost and that will be beneficial for society and the environment regardless of how climate change, policy changes, or other stresses unfold. Indeed, while our team did not specifically address the concept of *adaptive risk management* (briefly described in box 12.1, adapted from NRC 2010b), some of our approaches can be

BOX 12.1

COMPONENTS OF ADAPTIVE RISK MANAGEMENT (ADAPTED FROM NRC 2010).

Risk identification, assessment, and evaluation. Risks need to be evaluated by a range of affected stakeholders (who typically have different values and preferences) and by considering a range of factors. These include the impacts of allowing risks to go unmitigated, the costs of different risk management strategies, public perceptions and acceptability of risks and/or responses to them, as well as broader societal values that tend to favor certain general approaches to managing risk over others (e.g., a precautionary approach versus a cost-benefit or risk-benefit approach).

Iterative decision making and deliberate learning. Because many decisions will have to be made with incomplete information, and new information can be expected to become available over time (including information about the effectiveness of actions taken), decisions should be revisited, reassessed, and improved over time. This will require deliberate planning and processes for "learning by doing," as well as ongoing monitoring and assessment to evaluate both evolving risks and the effectiveness of responses.

Maximizing flexibility. Whenever decisions with long-term implications can be made incrementally (i.e., in small steps rather than all at once), the risk of making the "wrong" decision now can be reduced by keeping as many future options open as possible.

Maximizing robustness. When decisions have to be made all at once (for example, to build a piece of infrastructure or not), the risk of making the wrong decision can be reduced by selecting robust options—that is, options that maximize the probability of meeting identified goals and desirable outcomes while minimizing the probability of undesirable outcomes under a wide range of plausible future conditions.

Ensuring durability. Many policies need to remain in place, albeit in modified form, for many decades, in order to achieve their intended goals. This requires mechanisms that can ensure the long-term durability of policies and provide stability for investors and society while allowing for adaptive adjustments over time to take advantage of new information—a significant challenge for policy and institutional design.

A portfolio of approaches. In the face of complex problems, where surprises are expected and much is at stake, it would be unwise to rely on only one or a small number of actions to "solve" the problem without major side effects. A more robust approach would be to employ a portfolio of actions to increase the chance that at least one will succeed in reducing risk and to provide more options for future decision makers.

Effective communication. An essential component of effective risk management is the communication of risks, including the risks associated with different responses, to all involved stakeholders, including public, private, and civic sector stakeholders, as well as expert and lay individuals familiar with, or potentially affected by, the risks at hand.

Inclusive process. Since climate-related risks affect different regions, communities, and stakeholders in different ways and to different degrees, stakeholders should be included in significant roles throughout the process of identifying risks and response options, determining and evaluating what risks and responses are "acceptable" and "unacceptable," as well as in the communication and management of the risks themselves.

useful components of risk management under uncertainty. We believe this should be an operating principle of many agricultural systems in a world of change.

Changing the "agenda" to one that embraces sustainability objectives requires more than near-term economic incentives; voices that speak for an expanded set of interests must be heard.

While we did not study the psychology of decision making by growers in the valley, we did take away a sense that there are real barriers, at least for some decision makers, associated with the *idea* of change—whether it is toward the diversification of crops, more efficient fertilizer practice, changes in the overall level of government assistance, or recognition of environmental consequences. Economics is not all that matters; history, social relationships, and personal values all matter, and that means that change can be slow even under reasonable economic incentives. Through our knowledge system research, we became deeply aware of how challenging it is to change the agenda of a place, especially a place that is so subject to external forces. In the Yaqui Valley, we watched as the agenda shifted from one born in the *green revolution*—the encouragement of high yields and ever increasing inputs—to one that more closely resembles a sustainability agenda, in which increasing profits and reduction of negative externalities are both items of concern. Agendas are slow to change because they are set by those in power, those at the decision-making table. Agendas are reinforced because those decision makers determine what gets attention, and in some cases, what kind of research gets done. Voices change, and new voices are heard, but only slowly. Our voices, through our research funded by organizations inside and outside the region, mingled with others to influence the agenda of the Yaqui Valley to one that includes concern for off-site environmental and human health issues.

Knowledge matters, but so does connecting knowledge with action.

Sustainability approaches are often *knowledge intensive*, and when we could engage our research team in the development and application of knowledge for decision making, we usually made at least a small difference. But the range of questions was enormous, and new questions continuously emerged as we worked together. There were many, many areas in which we lacked the right expertise, or enough time and money, to be helpful (some of these are discussed later). For many of these questions, our team was insufficient, and other experts were hard to find or to coax into joining us. However, it was not just research expertise that we lacked—we sometimes lacked the time and mechanisms to connect what we were learning to decision making. We were extremely fortunate to have highly respected scientists (in agronomy and in water-resource management) on our team who worked directly with producers and other stakeholders to incorporate sustainability goals—they were boundary individuals, with one foot in the research world and one foot in the management world. To them we owe thanks for making sure that our research was relevant, useful, and actually used. However, we lacked people who could play that role at a different level—for example, to provide policy options to decision makers in Hermosillo (the state capital) and in Mexico City, and to help outline the types of incentives needed to diversify and stimulate agriculture wisely in the Yaqui Valley. We also lacked the time and expertise to fully engage and partner with relevant decision makers, and to make sure that we were communicating quickly and clearly what we had learned. All too often we found that we had engendered great interest in the results of our work but lacked the infrastructure to provide that knowledge in an ongoing way. Clearly, we needed additional partners at the table to help us do that, but in some areas in which we worked, such partners did not exist.

Finally, we learned that the links between researchers and decision makers are always changing—the knowledge system is dynamic. Relevant partners in linking knowledge to action changed over the time of our team effort (see chap. 5), and sometimes we were unaware of the ongoing changes. This point may be a crucial lesson for any attempt to construct mechanisms for linking knowledge and action, or for any research team that wishes to engage in that activity. We learned the drawbacks of making assumptions about the identity of our most critical actors or decision makers in the knowledge system; we learned not to assume, but to study and understand the players in the knowledge system, early and often, and actively identify key points where the knowledge to action links can most effectively take place.

Critical Missing Pieces in Studies of the Yaqui Valley Human-Environment Systems

As noted, our years of research raised as many questions that went unanswered as were addressed. We wished we could dedicate research effort to better understand, for example, knowledge barriers to diversification; the social consequences of constitutional reform for *ejido* communities' and Yaqui Amerindians' livelihoods, health, and welfare; the extent and ecosystems consequences of downwind transfers of pollutants from agriculture; the public health consequences of agriculture; the ecosystem services being provided by the streams and channels linking land to the oceans; the changing opportunities for marketing of commodities; the implications of future climate change, and many others. To address these critical questions about the human-environment system, we would have needed, along with more people and more expertise, a dramatic increase in information about population dynamics in the region, and additional and longer-term data bases on a range of critical measures and metrics. In the following paragraphs, we reflect on these two critical and understudied areas, with the hope that other research efforts, in the Yaqui Valley or elsewhere, will prioritize and incorporate them.

Population Dynamics in the Yaqui Valley

As our work evolved, it became increasingly clear that we needed an explicit focus on population dynamics if we were to succeed in understanding the social and human dynamics associated with land- and water-use decisions in this region.[1] We incorporated population data through two sources: census data and household surveys. Although census data provide general information such as age structure, gender, and household size, they lack detailed, household-level information about income, migration, and health of the population. Nonetheless, the census data show that urban populations are increasing and rural populations are remaining relatively stable at the municipality level. Moreover, throughout the Yaqui Valley, larger towns are growing faster than smaller ones, and in many cases, smaller localities are disappearing. We lack a nuanced understanding of why these changes are occurring and their direct role in land-use changes in the Yaqui Valley. Our household surveys provided detailed information about household characteristics, but those surveys were designed to collect information about different sectors of the population (e.g., aquaculture households, farmers), and not of the entire population. Accordingly, we were unable to extrapolate our survey data to characterize household dynamics for the entire region.

We also failed to spatially disaggregate data on migration. Such data can be obtained through household surveys or estimated via the *residual method*, whereby the rate of natural increase is estimated (births minus deaths) and then deducted from the total population increase. We hypothesized that many of the population changes have occurred within the region as a response to exogenous shocks. For example, it is likely that purposeful economic and technological development of the region during the early years of the green revolution encouraged in-migration. Over time, however, improving mechanization and increasing sizes of lands being managed by single owners may have resulted in surplus agricultural workers. A failure to diversify the Yaqui agriculture resulted in an increase in the area of grain crops such as wheat that are highly mechanized with a low labor requirement, as opposed to vegetable or fruit crops that require more labor. Whether these workers migrated out of the region or were absorbed by local urban or rural economies is unclear from the available census data. Ciudad Obregón and Navojoa experienced significant economic and urban growth over the last three decades, but population data aggregated at the city level make it difficult to answer questions about inter- and intracity socioeconomic differentiation and income and welfare dynamics. Furthermore, the census data classifies population based on residence; a household is labeled rural if the home structure is located in a nonurban place. As it becomes more common for rural residents to work in the city, this urban-rural dichotomy may not be sufficient for fully describing the demographic characteristics of the area nor for accounting the role of population on land-use change. How significant is rural nonfarm employment and incomes to rural households, and how has this contributed to out-migration? Likewise, have changes in Article 27 that allow *ejido* land to be rented and sold resulted in significant alteration of ownership patterns and out-migration, and if so, what are the consequences for land uses? And did these changes have indirect impacts on the Yaqui Amerindian communities, farming practices, or livelihoods? None of these questions were adequately answered in our study, but all would have improved our understanding and provided important insights into change in the valley.

Information on population and household dynamics are likewise important in the study of vulnerability. Our conceptual vulnerability framework (chap. 6) directs us to examine the access of individuals and groups to human and social resources as well as biophysical capital. We know that access to credit and information varies among communities and individuals, but we can only surmise that differential access is influenced by population characteristics and dynamics. Are different components of the population differentially at risk under external pressures resulting from policy changes or climate change? As the land tenure law changes lead to changes in ownership, will it decrease, or increase, the vulnerability of the *ejido* populations? Does the level of education, household size, age structure, or other population variables influence the ability of people to respond to external stresses? How have neighborhoods and place-based community ties influenced information sharing and risk management? While it is easy to state such questions, access to, and development of, appropriate data that allow their answers is more difficult and would have required focused, on-the-ground research at a level that our team was not able to engage in, because of lack of engaged expertise and lack of funding.

Adequate Temporal Databases

For a great many of the questions we asked, we needed long-term or at least multiyear data.[2] Very few research projects are sufficiently supported to collect and manage long-term monitoring data—ours certainly was not—yet such data are crucial to understanding change and influencing transitions. To meet some of our data needs, we took advantage of the years of work done at the International Maize and Wheat Improvement Center (CIMMYT) and other local research institutions. Data at the field and farm level on yields and inputs, farm budgets, and farmer perspectives were available for decades, and in comprehensive form since 1980, thanks to the work of CIMMYT. Soil maps and climate data were available from the National Institute of Forestry, Agriculture, and Livestock Research (INIFAP), National Water Commission (CNA), and Agricultural Research and Experimentation Board of the State of Sonora (PIEAES). Data on land and ocean characteristics were also available from remote sensing data from Thematic Mapper, SeaWiFS, AVHRR, and other sensors. As part of our projects, we obtained and compiled remote sensing databases going back several decades, and also obtained current data. We also used socioeconomic data from the National Institute of Statistics and Geography (INEGI) to provide a regular snapshot of demographic and economic trends. INEGI is an organization of the federal government of Mexico that is responsible for the collection of data on national accounts, prices, socioeconomics, and geography. Like most other government statistical bureaus, INEGI collects detailed census statistics every ten years, and at higher temporal frequency for select data. These data are available at multiple spatial scales, subdivided along administrative units (e.g., state, municipality, and locality). However, while these data provide information for a large area over a long time period, they are not available with enough temporal frequency to present a detailed picture of the dynamics of the system. Finally, we compiled and used water-resource databases, including input and outputs from reservoirs and spatially and temporally defined irrigation allocation values that had been collected by CNA. Beyond these, we had to collect our own data.

Over the course of our project we carried out several farm-level surveys on general agricultural conditions, with each survey focused on a particular topic: nitrogen fertilizer applications (1994, 1996, 2001, 2003), farm management practices (1996, 2001, 2003), and land ownership and rental agreements (1999, 2001, 2003). These surveys became part of a longer-run survey effort by CIMMYT to document the uptake of agricultural technology, resource management, and socioeconomic change in the Yaqui Valley. Using structured and in-depth interviews, we developed a rich dataset on farm characteristics, with much less information at the household level. The principal advantage with conducting surveys is the ability to acquire exact (or proxy) variables of interest (e.g., farm profits, household income, management practices). However, surveys are time and labor intensive, and their temporal and spatial extents are limited to the duration and budget, respectively, of the project. If the sample size is large or the distances between samples great, a survey can quickly become cost prohibitive.

In the course of our projects, we also collected biogeochemical and hydrologic data from field plants and soil, drainage canals, estuaries, groundwater wells, and air; these data were complemented by the longer-term measurements of climate variables and water use in the valley, but, for many of the questions that we wished to ask, these databases were insufficient. For example, we struggled to calculate valleywide nitrogen budgets, depending on our own several years of sampling of streams and canals, when what was required was a longer-term, more spatially complete, record of water flows and chemistry. Such data are often collected in the United

States and other parts of the world but were missing here. Likewise, understanding the downwind consequences of nitrogen oxide emissions from agriculture required good information on air quality in Ciudad Obregón and downwind deposition monitoring sites. These measurements, too, were not available and their absence made it impossible to characterize the full implications of overfertilization in the region, although back-of-the-envelope calculations suggested they could be large. Even for the sampling that we carried out ourselves, there were significant challenges and difficulties related to collecting the biophysical data, setting up and maintaining analytical instrumentation, transporting samples across international borders, and managing tremendous amounts of data being collected at fine temporal resolution and analyzed at different places.

Our surveys suggest that public health concerns may prove to be a stronger driving force in agricultural management and policy decisions than are environmental concerns. Researchers are beginning to link the burning of crop residues and emissions of nitrogen oxides to respiratory disease and other negative health impacts with heavy pesticide, herbicide, and fungicide exposure. Such health impacts of exposure are common in intensive agricultural farming communities throughout the developing world. In interviews, residents of the valley broadly claimed that the region suffers from high rates of leukemia and other cancers, but this effect has not been documented in peer-reviewed literature. Without good information documenting disease trends and monitoring pollutants from agriculture and other sources, it is difficult for health and environmental institutions to influence the agenda of the valley and to require changes in agriculture on behalf of those who may suffer from its impacts.

Reflections on Research Organization and Participation

As we carried out our research, we increasingly recognized that we were actors in a broader knowledge system for agriculture, water resources, and sustainable development. Many of our research team—students, academics, and research scientists from national and international research and management organizations—wanted to make their science useful to decision makers and thus intentionally engaged the valley's knowledge systems, which we purposely studied and participated in. At the same time, while many of us were fascinated by exploration and discovery in our own disciplinary areas, we were also engaged in the challenges of learning new disciplines and integrating them with our own. It was a complicated research environment. The experiences of the researchers, the challenges we faced, the way we organized, and the lessons we learned are explored in the following paragraphs.

In organizing our research, we started small and simple, and built to more complicated research networks over the years. Our early interactions were primarily among only three investigators—Matson (biogeochemist/environmental scientist), Naylor (economist), and Ortiz-Monasterio (agronomist)—and our research groups. Our research interests were driven by broader issues related to food security and global environmental change (see chap. 3); our research funding was from the USDA, Ford Foundation, the Pew Charitable Trusts, and Matson's MacArthur Fellowship. All three investigators were in it from the start, and the process of bringing our perspectives and research teams together to develop the research design was built on mutual respect and the lack of dominance of any one discipline over the others. A key element of our approach was to work with and for each other on the ground; for example, Naylor spent time in the lab and field learning firsthand about the biogeochemistry instrumentation and measurement, Matson went to the field with Ortiz-Monasterio and Naylor to understand the agronomic and economic questions being tested, and Ortiz-Monasterio became involved with all dimensions of field and lab work. It took time to learn and understand one another's perspectives and knowledge base, but the result was well worth it: *It is fair to say that the kinds of questions that we ultimately asked together, and the combination of experiments, measurements, and analyses that we made were different than any one of us would have made working alone.* This first project, motivated by a simple but nonetheless integrative interest, was the most easily funded piece of Yaqui Valley research. It probably would not have worked so well had we attempted to tackle something more complex.

It was impossible, however, to work in the Yaqui region and not notice the broader and more complex set of issues that were influencing decision makers—changing resource availability, information availability, global markets, national and international agricultural policy changes, and so on. Questions about water resources, climate change, macroeconomic upheaval, and national and international policy changes raised concerns about vulnerability. Rapid aquaculture and livestock expansion were hard to ignore and raised more questions about economics and environment. Changes in land tenure laws opened new questions about land-use patterns and population dynamics. Farmer associations and credit unions began to take on new roles. As new issues emerged, those of us who were in a position to notice them began a dialogue about what research could or should be done to address them, and what or who would be needed to do so. At our monthly or bimonthly team meetings (typically held at Stanford University, with members calling in where necessary) or annual all-hands meetings (typically held in the Yaqui Valley), we invited new investigators (both established and student investigators) with particular knowledge about the issues that were just coming onto our radar screen—agricultural and development policy experts, hydrologists, geographers, climate scientists, mathematical modelers, remote sensing experts, conservation scientists, sociologists, physical and biological oceanographers, and so on—and sometimes they chose to join in. With their engagement, we identified potential funding sources and wrote proposals, many of which did not succeed. We had big, integrative, multidisciplinary research ideas, and in the mid- to late 1990s there were not yet federal funding sources for them, nor were many reviewers willing to bet on our ability to work together in this way.[3] Typically, our research had to be funded in pieces, with each discipline bringing their own sources to the table, although seed money from Stanford's Institute of International Studies Bechtel Initiative funds, and later, a large grant from the Packard Foundation, helped us build the matrix into which all these pieces fit. As new funding could be patched together (and sometimes even without such support), new members were added to the whole. As our group and resources grew, we also increased our connections with regional NGOs and other researchers working in the broader Gulf of California area. Over time, influenced by our interactions with the international global change research programs and the nascent international sustainability science community, we began collaborations with separate interdisciplinary teams and comparisons with other places, using the Yaqui Valley as a case study for comparative analyses (see, for example, chaps. 5 and 6). While it would be ridiculous to say that we planned this long-term project, we did indeed do careful planning to bring each new piece into the whole.

In most of our research, our intent was to aid decision making in the Yaqui Valley, but also to address issues that were both scientifically interesting and useful in many other places around the world. Some of us came originally from a research world where rewards focus on the discovery of new knowledge; others were employed and rewarded for providing science in support of decision making. The Yaqui Valley allowed us to work in both these areas, often at the same time. In the midst of research on fertilizer management, for example, we discovered more about the role of nitrification in controlling the emission of nitrous oxide—a mechanism that had been recognized from the laboratory but not measured in the field, and that remained unaccounted for in greenhouse gas emissions models and reduction strategies. Likewise, the role of climate in wheat yield variation had never before been so clearly and forcefully shown as with our remote sensing–based time series analyses. Our in situ studies provided a textbook illustration of thermodynamic controls on trace gases from water systems, and synoptic, remotely sensed measures demonstrated relationships between phytoplankton dynamics in the ocean and activities on land for the first time. Concepts and analytical frameworks for vulnerability analysis emerged in part from our understanding of the situation in the valley, and hypotheses about the functioning of knowledge systems for sustainability were tested in our sites. And finally, the effects of major macroeconomic change (e.g., a sudden devaluation of the Mexican peso and sharply escalating interest rates that dominated farm production costs in the early to mid-1990s) provided a clear picture of how and why macro- and microeconomic evaluation should be integrated into any study of agricultural development. Our findings were presented in scientific journals and as invited and contributed presentations at many scientific meetings.

Much of what we learned was shared with decision makers, including farmers, credit union and farmer association leaders, civic leaders, water resource managers, and so on. As noted in chapters 3, 5, and 9, decision-maker participation in the research was a crucial form of shared learning, and on-farm trials of alternative management practices and new tools were important. Workshops, farm days, and other more formal communications were also used to engage decision makers in what was being learned and recommended. Research findings were published in a variety of CIMMYT and CNA white papers and handout materials. Remotely sensed images of crop yields, phytoplankton blooms in the Gulf of California, and aquaculture change were also shared and served as excellent boundary objects that engaged the research and decision-making communities together.

While overwhelmingly positive, this merging of science that could push the boundaries of knowledge and discovery with science in support of real-time decision making at times raised career concerns for members of the team. For example, in asking agronomic or water resource researchers to walk the fine line between developing and recommending new, innovative approaches versus providing dependable, low-risk (and often conservative) advice, were we exposing our colleagues to potential failure? By engendering interest in new approaches and tools, were we risking their ability to meet an ever-growing demand for the new information by their constituencies; were we promising more than we could deliver? And, as noted below, were we risking our student and junior faculty team members' academic research careers by engaging them in near-term problem solving research? Over time, we learned to walk that fine line together.

Much that we learned in the Yaqui Valley is transportable. At times we worried that our deep focus in this one place would help that one place, but not others. Indeed, one of the more common criticisms of our case study was that so much effort was being spent on one small place, when so much of the world needed similar kinds of work. With that concern in mind, we explicitly built or tested approaches, tools, metrics, and models that can be applied in other places. Moreover, many of our researchers are in positions that expect and even require them to carry the knowledge and perspectives to other regions of the world. Finally, we selected the Yaqui Valley in part because it is home to a major international research organization, CIMMYT, and regularly receives scientist and manager visitors from many other regions of the developing world. What we learned there had the potential of being shared more widely, and indeed, has been or is being transferred into policy and practice in Mexico and in other developing countries.

Perhaps as important, the Yaqui Project helped launched a new generation of researchers who understand what it means to be multi- or interdisciplinary in their perspectives or approaches, and who are committed to research for the welfare of people and the environment and resources on which they depend. This younger generation included both disciplinary scientists (e.g., biogeochemists, economists) as well as interdisciplinary scientists (e.g., those combining remote sensing and GIS-based analyses of land-use change with policy analysis), but all of them had the opportunity to gain multidisciplinary perspectives and engage with the larger group in asking interdisciplinary questions—questions that could not be asked by any one discipline working alone. Responsibility for the intellectual development and career paths of these young people sometimes weighed heavily on the older members, especially early in the project. Was this a fantastic opportunity or a career risk? Would this take too much time (given that interdisciplinary projects, in our experience, often take more time and effort than disciplinary ones, as do efforts that include linking knowledge with decision making)? Would they get jobs after this? With the passing of time, the proof was in the careers of these young scientists themselves, all of whom have gone on to other positions of their choice—in academia, consulting groups, government, foundations, and other nongovernmental organizations. For many of our students, the interactions and participation in the multidisciplinary team left a lasting mark that will influence their research and training in years to come.

Reflections on Impacts in Academic and Research Institutions

The Yaqui Project represented, for many of us, a first-time experience of working with researchers in disciplines or academic fields far from our own. Many of us had experience working in multidisciplinary teams, but those were by and large *near-field* multidisciplinary interactions; in the Yaqui Project, we were forced to deal with ideas, methods, and approaches very different than those we grew up with in science. We discovered the fun and exhilaration of working in this way, and also the frustrations and extra time taken in doing so. As noted earlier, some of us also experienced for the first time the challenges of linking what we were learning with decision making. As it turned out, the experimenting that we were doing as we progressed in our research programs yielded good things, not only in terms of contributions to science and to the Yaqui Valley and to other decision makers, but also to the development of interdisciplinary sustainability research at our home institutions. Moreover, the experience and learning within the Yaqui Project and through interactions across our many other research connections helped us articulate lessons that are relevant to research institutions more broadly.

Impacts in Our Home Institutions

At Stanford, our team's original home base was in the Institute for International Studies (now the Freeman Spogli Institute for International Studies [FSI]), an interdisciplinary policy institute led by team member Wally Falcon throughout the 1990s. The institute provided seed funding at critical times, and encouraged the merging of science and policy analyses that was the early hallmark of the Yaqui Project. Within several years after the launch of the project, and in part because of our team members' efforts, the interdisciplinary Center for Environmental Science and Policy was created within the institute; by the early 2000s, that center morphed into a much larger endeavor, now called the Woods Institute for the Environment, that brought together researchers from all seven schools of the university, explicitly encouraged a focus on interdisciplinary sustainability research, and explicitly incorporated the goal of linking knowledge to action. Seed funds launched a number of new interdisciplinary projects, and the institute fostered and supported the efforts of research teams to communicate and connect with decision makers. The Yaqui project was an early example of the kinds of research now engaged in widely at Stanford.

Among the most exciting of those projects is the Food Security and Environment (FSE) Program, a joint program of the FSI and Woods Institute. Team member Roz Naylor leads the interdisciplinary effort, engaging a new generation of students and many new faculty in efforts at the interface between food, hunger, and environment. FSE focuses on research in China, Indonesia, India, Brazil, Chile, and several countries in Sub-Saharan Africa—all of which play significant roles in the world food economy and are challenged by food security and environmental issues related to agriculture. The Yaqui Valley project illustrates the type of research conducted by FSE, with a focus on crops, livestock, aquaculture, biofuels, climate, water, and policy. Several of the researchers engaged in the Yaqui Project now work on other projects at FSE in different parts of the world.

The kinds of graduate students we were training, and the interdisciplinary nature of some of their research, also provided an early model for interdisciplinary graduate education at Stanford. While all of our students engaged in multidisciplinary interactions, a few of our graduate students learned and brought together multiple disciplines in their own research—they became true interdisciplinarians. Those who did so did it more or less in spite of their departmental degree programs, sometimes with their advisors running interference for them. The experiences of these students and their advisors contributed to the broader call for an interdisciplinary graduate program that crossed fields and departments. In 2000, our team members and other Stanford faculty developed a proposal for an interdisciplinary graduate program in environment and resources; within the proposal were several worked examples of the kinds of students, education, and research that would be expected within this program—and Yaqui Project students were the models. The Emmett Interdisciplinary Program in Environment and Resources (E-IPER) now is home to PhD students as well as MS students who are jointly obtaining JDs, MBAs, MDs or other graduate degrees. Many of them are carrying out research as part of interdisciplinary teams under the auspices of the Institute for the Environment.

Our experience with research and education for sustainability in the Yaqui Valley also became one proof-of-concept for Stanford's twenty-first-century initiatives to help solve problems and train leaders who are prepared to address the complex problems of the century (http://multi.stanford.edu/). Initiatives on environment and sustainability, human health, K–12 education, arts and creativity, and international democracy and development were all launched in the mid-2000s and are meant to foster the kinds of interdisciplinary interactions that the Yaqui Valley project illustrated. Indeed, the Yaqui Project was discussed and presented to illustrate what we *could* do if we decided to work together to address real-world problems.

Would these research and educational changes have occurred at Stanford without the Yaqui Project's experience? Probably so, but perhaps not so early or easily. The Yaqui Project provided evidence that this kind of research and education could be done well, could engage faculty and students alike, and could contribute both to improved understanding about how the world works and to problem solving.

At CIMMYT, our research brought greater awareness in the institution about the importance of considering and measuring the environmental impact of agronomic practices. In the area of fertilizer management, our evidence that improved management practices could lead to win-wins for farmers and the environment has led to similar efforts in other regions. In other areas of agronomic research at CIMMYT, including research on conservation agriculture, our work encouraged a stronger focus on the potential consequences of technology changes on resources and the environment. Measurement of environmental impacts from agricultural practices and technologies has become a part of CIMMYT's efforts in other areas in and outside Mexico, as has an interest in vulnerability and response to climate change. However, expanding research to include, for example, measurements of environmental impacts like greenhouse gas emissions, poses an important challenge in terms of logistics, equipment, and trained scientific staff in many developing countries, reinforcing the benefits of and need for interdisciplinary collaborations to work toward sustainability goals in other regions of Mexico and the world.

Lessons for Mobilizing Science and Technology for Sustainability

These reflections, and our experience with other groups trying to engage their research communities in sustainability science, suggest a number of things that are needed within institutions (especially academic institutions).[4] First, academic institutions need to find ways to facilitate interdisciplinary efforts that draw on the strengths of many different disciplines, allowing researchers to combine and integrate their knowledge around specific sustainability challenges. Working across disciplines takes time, respect, and openness to new ideas and new people; no one discipline can dominate. Working on sustainability challenges takes humbleness and openness to the knowledge and knowhow of people outside academia. Researchers working as part of interdisciplinary approaches to sustainability challenges often characterize this kind of research as the most fun thing they do, but it is often done outside the traditional research expectations of the university. Like Stanford, a number of universities now are engaged in experiments around the theme of encouraging such interactions. Some have identified new schools or colleges within the university with the explicit role of interdisciplinary problem solving. Others have developed umbrella institutions that are meant to harness the dispersed disciplinary and interdisciplinary strengths of the university and facilitate and incentivize research interactions that integrate them. Still others have instituted freestanding centers that operate more or less independently from the academic portions of the university. Some also provide seed money and incentive grants to entice and assist people as they work together for the first time. And some have employed more than one approach. We should actively learn from these experiments.

Recognition of interdisciplinary work is also key—both to successful scientific careers and to raising awareness of academic roles in the field. Having *Science*, *Nature*, or *Proceedings of the National Academy of Sciences* publish the science outcomes of this kind of effort raises recognition of both the field and the researchers (Clark 2007). Within universities and professional organizations, high-level attention to, and recognition of, researchers and their work (by university leaders, university press and outreach venues, and through professional awards, for example) are also very helpful.

Universities are also starting up programs for training students who understand and work within the broad context of sustainability, who sometimes carry the strengths of more than one discipline, and who can combine multiple disciplines, either themselves or through team efforts, to address questions that most of us more traditionally trained disciplinarians are challenged to do. Demand for such interdisciplinary programs is rapidly growing.

Much of sustainability science is fundamental research, but it is also use inspired and oriented toward decision making. Just as in the agricultural and medical fields, science-public interactions, outreach, and knowledge sharing are crucial aspects of sustainability science, yet most universities do not have well-honed mechanisms for the kind of dialogue, multidirectional information flow, and partnerships that are needed for sustainability science to be actually useful and used in decision making. Again, experiments are taking place with new kinds of research partnerships between academic, NGO, corporate, governmental, or community groups; dialogues and workshops with multiple stakeholders; communication strategies, and the development of in-house or external boundary organizations that purposefully link researchers and decision makers. Such efforts are exceptionally challenging, especially to universities, because they represent costs for which there are very few traditional sources of funds. Ultimately, the development of regional or local entities—sustainability resource and research centers—may help to provide integration and connection among research communities and local actors.

Finally, a sustainability transition will require changes in state, federal, and international research and development efforts, with more focused and coordinated approaches to fund both parts of the knowledge-to-action effort—both fundamental research and links to decision making. Institutional change is hard, but it is needed today at all levels if we are to successfully engage the science and technology community in a transition to sustainability.

Concluding Thoughts

The Yaqui Valley intensive agricultural region was born in the green revolution but is now evolving, ever so slowly, toward sustainable resource use and the *doubly green revolution*. Our study has been useful in terms of understanding the dynamics of human-environment systems in this place and time, and also in helping the region transition to sustainability. For all of us, the Yaqui Valley provided a real-life laboratory that allowed us to make serious contributions to scientific knowledge—and to problem solving. While this particular team effort is over, the research continues, and for the researchers who remain or will join in, there are many exciting challenges ahead.

Quite obviously, there's no one magic solution, no all-encompassing, win-win opportunity in agriculture, nor are there in water, energy, health, or other areas of sustainable development. But we can engage researchers and decision makers together in developing the knowledge, tools, and approaches that will allow agriculture and our other life-supporting endeavors to become more sustainable, and we can encourage governing institutions to support such changes. Large changes are needed, but small steps can make a difference in fostering a transition to sustainability, especially if they help shift the conversation and agenda toward the goals of sustainability.

NOTES

Chapter 2

1. Article 27 defines "non-irrigated equivalent" as follows: one hectare of irrigated land is assumed equal to two hectares of rain-fed agricultural land, four hectares of pasture land, and eight hectares of grazing land.

2. The Angostura, Oviachic, and El Novillo dams irrigate 83 percent of the land (Salinas-Zavala et al. 2006).

3. All water used for irrigation in the valley eventually drains into the Gulf of California through a system of five principle and fourteen smaller open canals (Harrison and Matson 2003).

4. Not all short-statured varieties enjoyed such a high yield. Many were released without extensive testing under pressure to keep ahead of the changing races of leaf and stem rust and in response to increased demands for various industrial quality wheats (Fischer and Wall 1976).

Chapter 5

1. Our analysis of knowledge systems in the Yaqui Valley was initiated through the NOAA-funded project title "Knowledge Systems for Sustainable Development" (KSSD). The multi-investigator, multisite project sought to understand and promote the design of effective systems to harness research-based knowledge in support of decisions bearing on the joint goals of human development and environmental stewardship—that is, on goals of sustainability (Clark et al. in preparation). The Yaqui Valley is one empirical case study within the KSSD project exploring efforts to harness research-based knowledge for sustainability over a range of issues, scales, and regions.

2. This chapter has considerable overlap with material previously published in E. B. McCullough and P. A. Matson, "The Evolution of the Knowledge System for Agricultural Development in the Yaqui Valley, Sonora, Mexico," in *Proceedings of the National Academy of Sciences of the United States of America*. 2011. doi:10.1073/pnas.1011602108.

3. The International Maize and Wheat Research Center (CIMMYT) is the flagship maize and wheat organization of the Consultative Group on International Agricultural Research (CGIAR).

4. There are three common models for land tenure in the Yaqui Valley: private ownership, smallholder peasant ownership (the *ejidal* sector), and Yaqui Amerindian colonies. *Ejidos* were established in Mexico after the Mexican Revolution in 1917, when the government expropriated vast amounts of land from large private landholders for peasant settlement. Reforms to Article 27 of the Mexican constitution in 1992 allowed *ejiditarios* to legally rent and sell their landholdings for the first time. Private landowners curb ownership ceilings by renting land from the other two types of owners; legalization of this practice in 1992 has fueled its growth. *Ejiditarios* own about 50 percent of the valley's land, but they actively cultivate less than 20 percent of it (Alianza Campesina del Noroeste 2003), citing difficulty in obtaining credit as a common reason for ceasing cultivation.

5. Producers with smaller operations usually receive loans from the federal government's public agricultural credit institution, *Financiera Rural*, which provides a more limited range of services. Several credit unions similar to those described cater specifically to smallholders in the valley.

6. According to farm surveys conducted by CIMMYT in the 1980s and 1990s, and subsequent surveys conducted by Stanford University in 1994–95, 1995–96, 2003–04, and 2004–05. Key informant interviews were conducted alongside surveys during 2003–04 and 2004–05.

7. According to survey data and key informant interviews.

8. Obtaining good information about weather patterns months in advance, or about soil nutrient availability in real time, is quite difficult. Work began on using ENSO predictions and other weather forecasting to provide information on seasonal weather patterns, but most efforts were focused on soil resources.

9. According to survey data.

Chapter 7

1. The World Wildlife Fund is in the process of developing a set of sustainable shrimp aquaculture standards (http://www.worldwildlife.org/what/globalmarkets/aquaculture/dialogues-shrimp.html) while the Global Aquaculture Alliance (http://www.gaalliance.org), a shrimp industry trade organization, has created a set of best aquaculture practices employed globally by its member companies.

2. World Wildlife Fund (http://www.worldwildlife.org/what/wherewework/gulfofca/index.html); The Nature Conservancy (http://www.nature.org/ourinitiatives/regions/northamerica/mexico/placesweprotect/the-baja-peninsula.xml/); Environmental Defense Fund (http://www.edf.org/page.cfm?tagID=46066).

3. Monterey Bay Aquarium seafood watch (http://www.montereybayaquarium.org/cr/seafoodwatch.aspx).

Chapter 8

1. A complete listing of all acronyms can be found at the beginning of the book.

2. Financial profitability, also called private profitability, is equal to revenues minus costs at market prices. Economic profitability, also called social profitability, is equal to revenues minus costs calculated with prices adjusted for policy distortions, market failures, and environmental externalities. In this analysis, we quantify economic prices accounting for policy distortions only, as explained later in the chapter.

3. Five field surveys were completed under the auspices of Stanford University. All of the surveys dealt with general farm conditions and management practices, but each had a particular focus: nitrogen fertilizer (1994–96); farm management practices (1996); land ownership and rental arrangements (1999); aquaculture production and land ownership (2001); and water utilization under drought conditions (2003–04). For a description of methods used in each survey, see (respectively) Matson et al. (1998), Harris (1996), Avalos-Sartorio (1997), Lewis (2002), Luers (2004), and McCullough (2005). Hereafter, these field inquiries are referred to as the Stanford University surveys. References are also made to farm surveys on production technology and practices by CIMMYT's Economics Program, reported in Pingali and Flores (1998) and Meisner et al. (1992).

4. M$ refers to Mexican pesos and US$ to United States dollars. Unless otherwise noted, translations between M$ and US$ are made with respect to the average exchange rate for the period being cited. In 1991, for example, the nominal M$/US$ rate was 3.0; in 1996, the rate was 7.6; in 1999, the rate was 9.6; and in 2007, the rate was 11.0. (Refer to table 8.1 for a complete list).

5. The producer subsidy equivalent (PSE) is defined as the value of the subsidy needed to make domestic production of a commodity competitive in world markets. Producer subsidy equivalents are often given as a percentage of a commodity's value and reflect the share of farm earnings attributable to subsidies. For more discussion on the PSE in Mexican agriculture and its value relative to OECD countries, see Wise (2004). Similarly, the consumer subsidy equivalent is the amount or percent of subsidy consumers receive as compared to world prices.

6. J. E. Taylor and his colleagues have also tried to disentangle NAFTA effects from other policy reforms for Mexico as a whole. See, for example, Yúnez-Naude and Taylor (2006), Taylor (2002), and Yúnez-Naude (1998 and 2002). A very readable critique of analyses that wrongly ascribe actions and outcomes to NAFTA can be found in Krugman (1993).

7. By 1988, production subsidies through CONASUPO totaled more than US$500 million. In that same year, CONASUPO consumer subsidies also added nearly US$1 billion in budgetary costs (OECD 1997). An analysis by Avalos-Sartorio (2006) shows that during the last decade of CONASUPO benefits to maize farmers came overwhelmingly from raising the average commodity price, not from reducing price instability.

8. Throughout the first half of the decade, CONASUPO maintained producer price supports for maize and beans, plus consumer subsidies for maize tortillas (which were removed in 1998). Early on, ASERCA provided payments to wheat millers based on the volume of domestic purchases and sales; these payments ended in 1995 after consumer price ceilings for wheat flour and bread had been removed (OECD 1997).

9. The main institutions operating in this capacity included BANRURAL (National Rural Credit Bank), FIRA (Trust Fund for Agriculture, a secondary bank that discounts loans to commercial banks), FIRCO (Trust Fund for Shared Risk, operating through the Department of Agriculture), and FONAES (National Fund for Social Enterprises, operating through the Department of Social Development). The latter two promoted credit at zero interest (Puente-González 1999).

10. In 1991, FONAES began operation and has since become a principal lender of credit (Puente-González 1999).

11. Each irrigation module oversees its own budget for operation, conservation, and administration of secondary canals.

12. The M$ was fixed to the US$ for most of 1988, then allowed to depreciate the following three years at a preannounced rate. In November 1991, a publicly announced intervention band for the peso was established, with a constant floor and a ceiling that depreciated at a predetermined rate. The band was initially set at 1.5 percent and rose to 9 percent by the end of 1993 (Savastano et al. 1995).

13. As foreign investors grew wary of the administration during the course of 1994, they switched almost entirely out of *cetes* and into *tesobonos*—the newly issued, dollar-denominated, short-term bonds created to hold investor interest. During the course of 1994, the value of *tesobonos* jumped from US$1.4 billion to US$30.4 billion, while the value of *cetes* fell from US$22.7 billion to US$5.4 billion. This shift greatly increased the level of dollar-indexed government debt (Ramirez 1996).

14. Investor panic began shortly after President Zedillo was sworn in on December 1, 1994. By December 20, international reserves had fallen to US$10.5 billion and the exchange rate band was widened from 9 percent to 15 percent. Two days later, reserves fell by another US$4 billion and the peso was allowed to float (Savastano et al. 1995).

15. Locally known as the *Asociación de Organismos de Agricultores del Sur de Sonora* (AOASS).

16. See note 3.

17. This support was given through the Integral Agrarian Program of Sonora (PAIS), which established the first aquaculture *ejidos* in the region.

18. Of the 55 percent farmed by *ejidos*, 19 percent was farmed by collective *ejidos* and 36 percent by individual *ejidos*, in which commonly owned land had been parceled out to individual *ejidatarios*.

19. The Mexican GDP deflator was used to obtain real prices.

20. The shadow exchange rate is based on purchasing power parity; that is, the *real* exchange rate when corrected for relative inflation rates vis-à-vis Mexico's

main trading partners. Many additional assumptions were needed in the calculations leading to figure 8.4b, and these are described in Puente-González (1999).

21. Soybeans are not shown in figures 8.4a and 8.4b because the whitefly wiped the crop out of the valley shortly after 1991.

22. Regulations of pollution to water are included in the 1998 Ley Federal de Derechos en Materia de Agua (Federal Law Concerning Water Rights). Regulations covering pesticide use are in the 1996 revision of the Ley General del Equilibrio Ecológico y Protección al Ambiente (General Law of Ecological Balance and Environmental Protection). Regulations covering the burning of straw are covered in articles of Reglamento en Materia de Preservación, Conservación y Restauración del Equilibrio Ecológico y el Mejoramiento del Ambiente para el Municipio de Cajeme, 1997–2000.

23. These regulations are under the Comisión Federal para la Protección contra Riesgos Sanitarios (Federal Commission for Protection Against Health Risks).

24. Some progressive producers in the Yaqui Valley are adopting technology to redistribute livestock waste as fertilizer on nonfood crops in anticipation of the compliance targets.

25. This point became clear in a lengthy discussion with Luis Signoret, president of the Association of Producer Organizations in Southern Sonora (AOASS), in November 2006. Stanford surveys from earlier periods also suggested this view. Most farmers were reluctant to try high-valued crops that they knew little about in terms of the biophysical constraints and market potential.

Chapter 9

1. The on-farm experiments tested five rates of N: 0, 75, 150, 225, and 300 kg N/ha. All these rates were evaluated with two different timings of N application; one where 75 percent of the total rate was applied preplant, none applied at planting, and the other 25 percent at the time of the first postplant irrigation. This treatment represented the most common timing used by farmers. The other treatment was no N applied preplant, 33 percent applied one day before planting, and the other 67 percent applied at the time of the first postplant irrigation: this represented the timing of our best alternative. There were two additional treatments at the rate of 225 kg N/ha when all the N was applied either in the preplant application or at planting. Although most farmers use anhydrous ammonia during the first postplant irrigation, because of logistical problems it was impossible to use it, and urea was used for all the N applications, preplant or post.

2. Exchange rate 9.2 pesos/US$.

3. Exchange rate 7.6 pesos/US$.

4. UAN 32 percent solution is a nitrogen fertilizer composed of urea and ammonium nitrate. Approximately one-half of the N is from urea, and one-half is from ammonium nitrate.

Chapter 10

1. Also known as the Haber-Bosch process. For more information, see Vaclav Smil's book *Enriching the Earth* for an entertaining history of the hunt for cheap sources of fertilizer N and the discovery of the Haber-Bosch process (Smil 2001).

2. Typical management practices involve three-quarters of the ~260 kg N applied as broadcast urea approximately a month before planting, and the remainder applied as urea or anhydrous ammonia during the first postplanting irrigation. See chapter 3 for a thorough discussion on management practices in the valley.

3. Seven trials with differing management practices were evaluated for agronomic, economic, and environmental performance based on factors including grain quality, yield, estimated profit, and nitrogen losses. The best practice minimized nitrogen losses, maximized farmer profit, and maintained grain yield and quality.

4. Nitrogen typically weighs 14 g/mol, but a less common isotope of N weighs 15 g/mol (^{15}N). The scarcity of the heavier form of N makes it possible to trace as it changes forms in the soil, and N "tracer" studies add elevated concentrations of ^{15}N in one form and measure the increase or decrease of the heavier isotope as it is transformed in the soil.

5. Clay "fixation" should not be confused with biological N fixation, a microbially mediated process that converts atmospheric N_2 to biologically active forms of N. Clay fixation works in quite the opposite way, removing NH_4^+ from the biologically active pool of N.

6. There is a large amount of uncertainty surrounding these calculations, which are based on findings from controlled laboratory measurements under ideal conditions for denitrification. We also assumed a 1 cm depth of active denitrification, an estimate that is similar to the zone of denitrification determined in a range of estuarine sediments (e.g., Jenkins and Kemp 1984).

7. Using estimates of North American hog slaughter weights (95 kg) and ages (~ six months), the density of pigs in the valley is ~2.5 head/ha. This falls between reports of 1.6 pigs/ha of farmland in Iowa and more dense regions in Germany and the Netherlands containing 3.6 and 21.5 head/ha, respectively (Smil 2002).

8. Based on the difference between N in inlet and outlet water by Páez-Osuna et al. (1997) estimated to be 26.8 kg N/ha, equivalent to 236 Mg N for southern Sonora's 8,800 hectares in active production in 2004.

9. Farmer adoption of management practices that we identified as *best practices* has been slow, and a combination of farmer surveys, remote sensing, and economic analyses suggest that there are rational reasons for farmers to resist such a change (see chap. 3).

Chapter 11

1. Just before planting permits were authorized, a group of innovative producers appealed to the Yaqui Irrigation District with experimental data showing that wheat yields would not suffer if three on-plant irrigations were applied instead of

the customary three (chap. 9). The irrigation district decided to expand the planting area. They planned to meet irrigation demand by delivering one less irrigation to wheat fields and by relying on in-season runoff into the reservoir.

2. This estimate was made by summing the available storage under confined conditions (above the top of the screened wells), assuming a specific storage of 10^{-4}/m, plus the remaining storage under unconfined conditions, assuming a specific yield of 0.2 and counting two-thirds of the aquifer thickness.

Chapter 12

1. This section draws in part on material previously published in P. Matson, A. Luers, K. Seto, R. Naylor, and I. Ortiz-Monasterio, "People, Land Use and Environment in the Yaqui Valley, Sonora, Mexico," in *Population, Land Use, and Environment*, ed. B. Entwisle and P. Stern, 238–64 (Washington, DC: National Research Council, 2005).

2. Ibid.

3. This section draws on a previously published manuscript: P. A. Matson, "The Sustainability Transition," *Issues in Science and Technology* 25(4 [2009]): 39–42. Washington, DC: The National Academies Press.

4. Ibid.

REFERENCES

Aburto-Oropeza, O., E. Ezcurra, G. Danemann, V. Valdez, J. Murray, and E. Sala. 2008. "Mangroves in the Gulf of California Increase Fishery Yields." *Proceedings of the National Academy of Sciences* 105(30): 10456–10459. doi: 10.1073/pnas.0804601105.

Addams, L. 2004. "Water Resource Policy Evaluation Using a Combined Hydrological Economic-agronomic Modeling Framework: Yaqui Valley, Sonora, Mexico." PhD thesis. Palo Alto: Stanford University.

——. 2005. "Evaluating Increased Groundwater Use in the Yaqui Valley, Mexico." *Southwest Hydrology* 4: 8–9.

Aceves-Medina, G., S. P. A. Jiménez-Rosenberg, A. Hinojosa-Medina, R. Funes-Rodríguez, R. J. Saldierna-Martínez, and P. E. Smith. 2004. "Fish Larvae Assemblages in the Gulf of California." *Journal of Fish Biology* 65(3): 832–47.

Adger, W. N., S. Dessai, M. Goulden, M. Hulme, I. Lorenzoni, D. R. Nelson, L. O. Naess, J. Wolf, and A. Wreford. 2009. "Are There Social Limits to Adaptation to Climate Change? *Climatic Change* 93(3–4): 335–54

Adger, W. N., H. Eakin, and A. Winkels. 2009. "Nested and Teleconnected Vulnerabilities to Environmental Change." *Frontiers in Ecology and the Environment* 7(3): 150–57.

Ahrens, T. D. 2009. "Improving Regional Nitrogen Use Efficiency: Opportunities and Constraints." PhD thesis. Palo Alto: Stanford University.

Ahrens, T.D., J. M. Beman, J. A. Harrison, P. K. Jewett, and P. A. Matson. 2008. "A Synthesis of Nitrogen Transformations and Transfers from Land to the Sea in the Yaqui Valley Agricultural Region of Northwest Mexico." *Water Resources Research* 44, W00A05. doi:10.1029/2007WR006661.

Albert, L. A. 1996. "Persistent Pesticides in Mexico." *Reviews of Environmental Contamination & Toxicology* 147: 1–44.

Alianza Campesina del Noroeste. 2003. *Fondo de Apoyo para el Rescate de Parcelas del Sector Social*. Alianza Campesina del Noroeste, Ciudad Obregón, Mexico.

Alvarez-Borrego, S., J. Rivera, G. Gaxiola-Castro, M. Acosta-Ruiz, and R. Schwartzlose. 1978. "Nutrientes in el Golfo de California." *Ciencias Marinas* 5: 53–71.

Anderson, D., P. Glibert, and J. Burkholder. 2002. "Harmful Algal Blooms and Eutrophication: Nutrient Sources, Composition, and Consequences." *Estuaries* 25: 704–26.

Anderson, W. B., and G. A. Polis. 1998. "Marine Subsidies of Island Communities in the Gulf of California: Evidence from Stable Carbon and Nitrogen Isotopes." *Oikos* 81(1): 75–80.

Angelsen, A., and D. Kaimowitz, eds. 2001. *Agricultural Technologies and Tropical Deforestation*. Wallingford, United Kingdom: CAB International.

Aquino, P. 1998. *The Adoption of Bed Planting of Wheat in the Yaqui Valley, Sonora, Mexico*. Mexico, DF: CIMMYT.

Asner, G. P., S. Archer, R. F. Hughes, R. J. Ansley, and C. A. Wessman. 2003. "Net Changes in Regional Woody Vegetation Cover and Carbon Storage in Texas Drylands, 1937–1999." *Global Change Biology* 9(3): 316–35.

Avalos-Sartorio, B. 1997. "Modeling Fertilization Practices of Wheat Farmers in Mexico's Yaqui Valley." PhD thesis. Palo Alto: Stanford University.

——. 2006. "What Can We Learn from Past Price Stablilization Policies and Market Reform in Mexico?" *Food Policy* 31: 313–27.

Babcock, B. A. 1992. "The Effects of Uncertainty on Optimal Nitrogen Applications." *Review of Agricultural Economics* 14: 271–80.

Beman, J., K. Arrigo, and P. Matson. 2005. "Agricultural Runoff Fuels Large Phytoplankton Blooms in Vulnerable Areas of the Ocean." *Nature* 434: 211–14.

Beman, J. M., B. N. Popp, and C. A. Francis. 2008. "Molecular and Biogeochemical Evidence for Ammonia Oxidation by Marine Crenarchaeota in the Gulf of California." *ISME Journal* 2(4): 429–41.

Bograd, S. J., C. G. Castro, E. D. Lorenzo, D. M. Palacios, H. Bailey, W. F. Gilly, and F. P. Chavez. 2008. "Oxygen Declines and the shoaling of the Hypoxic Boundary in the California Current." *Geophysical Research Letters* 35: L12607.

Bowen, T. 2004. "Archaeology, Biology and Conservation on Islands in the Gulf of California." *Environmental Conservation* 31(03): 199–206.

Brady, N., and R. Weil. 1996. *The Nature and Properties of Soils*. Upper Saddle River, NJ: Prentice Hall.

Bredehoeft, J., and R. Young. 1983. "Conjunctive Use of Groundwater and Surface Water for Irrigated Agriculture: Risk Aversion." *Water Resources Research* 19(5): 1111–21.

Brinton, E., and A. W. Townsend. 1980. "Euphausiids in the Gulf of California–The 1957 Cruises." *CalCOFI Reports* 21: 211–36.

Bruinsma, J. 2009. "*The Resource Outlook to 2050: By How Much Do Land, Water and Crop Yields Need To Increase By 2050*" Paper presented at FAO (Food and Agriculture Organization) Expert Meeting, June 24–26, Rome, Italy, on "How to Feed the World in 2050."

Camacho-Casas, M. A. 2003. *Júpare C2001: Nueva Variedad de Trigo Cristalino Adoptada por los Productores del Noroeste de México.* Cd. Obregón, Mexico, DF: INIFAP.

Carpenter, S., N. Caraco, D. Correll, R. Howarth, A. Sharpley, and V. Smith. 1998. "Nonpoint Pollution of Surface Waters with Phosphorus and Nitrogen." *Ecological Applications* 8: 559–68.

Carpenter, S. R., and L. H. Gunderson. 2001. "Coping with Collapse: Ecological and Social Dynamics in Ecosystem Management." *Bioscience* 51(6): 451–57.

Cash D. W., W. C. Clark, F. Alcock, N. M. Dickson, N. Eckley, D. H. Guston, J. Jager, and R. B. Mitchell. 2003. "Knowledge Systems for Sustainable Development." *Proceedings of the National Academy of Sciences of the United States of America* 100(14): 8086–91.

Cassman, K. G., A. Dobermann, D. T. Walters, and H. Yang. 2003. "Meeting Cereal Demand While Protecting Natural Resources and Improving Environmental Quality. *Annual Review of Environment and Resources* 28: 315–58.

Centro Internacional de Mejoramiento de Maíz y Trigo (CIMMYT). www.cimmyt.org (accessed 2006).

Chameides, W. L., P. S. Kasibhatla, J. Yienger, and H. Levy. 1994. "Growth of Continental-Scale Metro-Agro-Plexes, Regional Ozone Pollution, and World Food Production." *Science* 264: 74–77.

Chan, F., J. A. Barth, J. Lubchenco, A. Kirincich, H. Weeks, W. T. Peterson, and B. A. Menge. 2008. "Emergence of Anoxia in the California Current Large Marine Ecosystem." *Science* 319(5865): 920. doi:10.1126/science.1149016.

Chen, Xin-Ping, Zhen-Ling Cui, P. M. Vitousek, K. G. Cassman, P. A. Matson, Jin-Shun Bai, Quin-Feng Meng et al. 2011. "Integrated Soil-crop System Management for Food Security." *Proceedings of the National Academy of Sciences of the United States of America*. 108(16): 6399–404.

Clark, W. C. 2007. "Sustainability Science: A Room of its Own." *Proceedings of the National Academy of Sciences of the United States of America*. 104(6): 1737–38.

Clark, W., J. Jager, R. Corell, J. X. Kasperson, J. J. McCarthy, D. Cash, S. J. Cohen, et al. 2000. "*Assessing Vulnerability to Global Environmental Risks*." Global Environmental Assessment (GEA) Working Paper.

Clark, W. C., et al. (in prep) *Proceedings of the National Academy of Sciences of the United States of America*.

Comisión Nacional del Agua (CNA). 1992. *Ley de Aguas Nacionales*. Mexico, DF: Comisión Nacional del Agua.

———. 1998. "*Clasificación de la Propiedad Agrícola de los Usuarios Ejidales y Números Ejidales en el Distrito de Riego No. 41, Río Yaqui, Sonora*." Working Paper. Cd. Obregón, Mexico: CNA.

Conway, G. 1997. *The Doubly Green Revolution: Food for All in the 21st Century*. London, UK: Penguin.

Cristiani, B.C. 1984. *Hoy Luchamos por la Tierra*. Autónoma Metropolitana Xochimilco: Mexico Universidad.

Cruz-Colin, E. In preparation. Bathymetry and circulation in Bahía del Tóbari.

Dabdoub, C. 1980. *Breve Historia del Valle del Yaqui*. Mexico City: Editores Asociades Mexicana, S.A.

Diaz, R. 2001. "Overview of Hypoxia around the World." *Journal of Environmental Quality* 30: 275–81.

Diaz, R., and R. Rosenberg. 1995. "Marine Benthic Hypoxia: A Review of Its Ecological Effects and the Behavioral Responses of Benthic Macrofauna." *Oceanography and Marine Biology: an Annual Review* 33: 245–303.

Diaz, S. C., M. D. Therrell, D. W. Stahle, and M. K. Cleaveland. 2002. "Chihuahua (Mexico) Winter-Spring Precipitation Reconstructed from Tree-rings, 1647–1992." *Climate Research* 22(3): 237–44.

Distrito de Riego del Río Yaqui. 2003. *Proyecto de Mejoramiento del Distrito de Riego del Río Yaqui, S. de RL de I.Y. Y C.V.* Ciudad Obregón, Sonora, Mexico: Distrito de Riego del Río Yaqui.

Dixon, J., A. Gullivar, and D. Gibbon. 2001. *Farming Systems and Poverty: Improving Farmer Livelihoods in a Changing World*. Washington, DC: World Bank.

Duce, R. A., J. LaRoche, K. Altieri, K. R. Arrigo, A. R. Baker, D. G. Capone, S. Cornell, et al. 2008. "Impacts of Atmospheric Anthropogenic Nitrogen on the Open Ocean." *Science* 320(5878): 893–97. doi:10.1126/science.1150369.

Eakin, H., and A. L. Luers. 2006. "Assessing the Vulnerability of Social-Environmental Systems." *Annual Review of Environment and Resources* 31: 365–94.

Economist. 2006. "A Survey of Mexico." November 18. http://www.economist.com/surveys (accessed 2007).

Federoff, N. V., D. S. Battisti, R. N. Beachy, P. J. M. Cooper, D. A. Fischhoff, C. N. Hodges, and V. C. Knauf. 2010. "Radically Rethinking Agriculture for the 21st Century." *Science* 327(5967): 833–34.

Felger, R. S., M. B. Johnson, and M. F. Wilson. 2001. *The Trees of Sonora, Mexico*. New York: Oxford University Press.

Feller, I. C., C. E. Lovelock, U. Berger, K. L. McKee, S. B. Joye, and M. C. Ball. 2010. "Biocomplexity in Mangrove Ecosystems." *Annual Review of Marine Science* 2(1): 395–417.

Finlay, K.W. 1968. "The Significance of Adaptation in Wheat Breeding." *Proceedings of the 3rd International Wheat Genetics Symposium*, 403–9. Canberra: Australian Academy of Science.

Fischer, R. A., and P. C. Wall. 1976. "Wheat Breeding in Mexico and Yield Increases." *Journal of the Australian Institute of Agricultural Science* 42(3): 139–48.

Flores-Verdugo, F., F. Gonzalez-Farias, D. S. Zamorano, and P. Ramirez-Garcia. 1992. "Mangrove Ecosystems of the Pacific Coast of Mexico: Distribution, Structure, Litterfall, and Detritus Dynamics." In *Coastal Plant Communities of Latin America*, U. Seeliger, ed., 269–88. San Diego: Academic Press.

Foley, J. A., R. DeFries, G. P. Asner, C. Barford, G. Bonan, S. R. Carpenter, F. S. Chapin, et al. 2005. "Global Consequences of Land Use." *Science* 309(5734): 570–74.

Food and Agriculture Organization of the United Nations (FAO). 1952, 1962, 1972, 1982, 1992, 1994. *Production Yearbooks.* Rome, Italy: FAO.

———. 2006a. *World Agriculture towards 2030/2050*. Rome, Italy: FAO.

———. 2006b. FAO Statistical Databases. http://faostat.fao.org (accessed 2006).

Freebarin, D. K. 1963. "Relative Production Efficiency between Tenure Classes in the Yaqui Valley, Sonora, Mexico." *Journal of Farm Economics* 45(5): 1150–60.

Galindo-Bect, M. S., E. P. Glenn, H. M. Page, K. Fitzsimmons, L. A. Galindo-Bect, J. M. Hernández-Ayón, R. L. Petty, J. García-Hernández, and D. Moore. 2000. "Penaeid Shrimp Landings in the Upper Gulf of California in Relation to Colorado River Freshwater Discharge." *Fishery Bulletin* 98: 222–25.

Galloway, J., F. Dentener, D. Capone, E. Boyer, R. Howarth, S. Seitzinger, G. Asner, et al. 2004. "Nitrogen Cycles: Past, Present, and Future." *Biogeochemistry* 70: 153–226.

Galloway, J. N., A. R. Townsend, J. W. Erisman, M. Bekunda, Z. Cai, J. R. Freney, L. A. Martinelli, S. P. Seitzinger, and M. A. Sutton. 2008. "Transformation of the Nitrogen Cycle: Recent Trends, Questions, and Potential Solutions." *Science* 320: 889–92.

García-Hernández, J., K. A. King, A. L. Velasco, E. Shumilin, M. A. Mora, and E. P. Glenn. 2001. "Selenium, Selected Inorganic Elements, and Organochlorine Pesticides in Bottom Material and Biota from the Colorado River Delta." *Journal of Arid Environments* 49: 65–89.

Gates, M. 1993. *Peasants, the Debt Crisis, and the Agricultural Challenge of Mexico*. Latin America Perspective Series. Boulder, CO: Westview Press.

Gebbers, R., and V. I. Adamchuk. 2010. "Precision Agriculture and Food Security." *Science* 327(5967): 828–31.

Gentry, H. S. 1942. *Río Mayo Plants: A Study of the Flora and Vegetation of the Valley of the Río Mayo, Sonora*. Washington, DC: Carnegie Institution of Washington Publication 527.

Gilbert, J. Y., and W. E. Allen. 1943. "The Phytoplankton of the Gulf of California Obtained by the 'E.W. Scripps' in 1939 and 1940." *Journal of Marine Research* 5: 89–110.

Glibert, P., J. Harrison, C. Heil, and S. Seitzinger. 2006. "Escalating Worldwide Use of Urea: A Global Change Contributing to Coastal Eutrophication." *Biogeochemistry* 77: 441–63.

Gilly, W. F., U. Markaida, C. H. Baxter, B. A. Block, A. Boustany, L. Zeidberg, K. Reisenbichler, B. Robison, G. Bazzino, and C. Salinas. 2006. "Vertical and Horizontal Migrations by the Jumbo Squid *Dosidicus gigas* Revealed by Electronic Tagging." *Marine Ecology Progress Series* 324: 1–17.

Godfray, H. C. J., J. R. Beddington, I. R. Crute, L. Haddad, D. Lawrence, J. F. Muir, J. Petty, S. Robinson, S. M. Thomas, and C. Toulmin. 2010. "Food Security: The Challenge of Feeding 9 Billion People." *Science* 327(5967): 812–18.

González, R., and L. E. Marín. 2000. "Modelo Hidrogeológico Conceptual del Acuífero del Valle del Yaqui, Sonora en un Contexto Geológico Regional." In *ITSON-DIEP Investigaciones*, 69–83.

González, R., L. E. Marín, and G. Córdova. 1997. Hydrogeology and Groundwater Pollution of Yaqui Valley, Sonora, Mexico. *Geofísica Internacional* 36:49–54.

Grantham, B. A., F. Chan, K. J. Nielsen, D. S. Fox, J. A. Barth, A. Huyer, J. Lubchenco, and B. A. Menge. 2004. "Upwelling-driven Nearshore Hypoxia

Signals Ecosystem and Oceanographic Changes in the Northeast Pacific." *Nature* 429(6993): 749–54.

Guston, D. H. 1999. "Stabilizing the Boundary between US Politics and Science: The Role of the Office of Technology Transfer." *Social Studies of Science* 29(1): 87–111.

Halpern, B. S., S. Walbridge, K. A. Selkoe, C. V. Kappel, F. Micheli, C. D'Agrosa, J. F. Bruno. 2008. "A Global Map of Human Impact on Marine Ecosystems." *Science* 319(5865): 948–52. doi:10.1126/science.1149345.

Harris, J. 1996. "Conservation Tillage: A Viable Solution for Sustainable Agriculture in the Yaqui Valley, Mexico." Senior honors thesis. Palo Alto: Stanford University.

Harrison, J., and P. A. Matson. 2003. "Patterns and Controls of Nitrous Oxide Emissions from Waters Draining a Subtropical Agricultural Valley." *Global Biogeochemical Cycles* 17(3): 1080. doi:10.1029/2002GB001991.

Harrison, J., P. Matson, and S. Fendorf. 2005. "Effects of a Diel Oxygen Cycle on Nitrogen Transformations and Greenhouse Gas Emission in a Eutrophied, Subtropical Stream." *Aquatic Sciences* 67: 308–15.

Hazell, P. B. R. 2009. *The Asian Green Revolution*. IFPRI Discussion Paper 911. Washington, DC: IFPRI.

Healy, E. C., R. Guadarrama, J. C. Ramirez. 1988. *Historia Contemporánea de Sonora, 1929–1984*. Hermosillo, Sonora, Mexico: Gobierno del Estado de Sonora.

Hernández-Ayón, J. M., M. S. Galindo-Bect, B. P. Flores-Báez, and S. Alvarez-Borrego. 1993. "Nutrient Concentrations Are High in the Turbid Waters of the Colorado River Delta." *Estuarine, Coastal and Shelf Science* 37: 593–602.

Hicks, W. 1967. "Agricultural Development in Northern Mexico, 1940–1960." *Land Economics* 43(4): 393–402.

Holland, E. A., B. H. Braswell, J. Sulzman, and J. F. Lamarque. 2005. "Nitrogen Deposition onto the United States and Western Europe: Synthesis of Observations and Models." *Ecological Applications* 15(1): 38–57.

Holtgrieve, G. W., D. E. Schindler, and P. K. Jewett. 2009. "Large Predators and Biogeochemical Hotspots: Brown Bear (*Ursus arctos*) Predation on Salmon Alters Nitrogen Cycling in Riparian Soils." *Ecological Research* 24(5): 1125–35.

Howarth, R. W. 2008. "Coastal Nitrogen Pollution: A Review of Sources and Trends Globally and Regionally." *Harmful Algae* 8: 14–20.

Howarth, R. W., A. Sharpley, and D. Walker. 2002. "Sources of Nutrient Pollution to Coastal Waters in the United States: Implications for Achieving Coastal Water Quality Goals." *Estuaries* 25(4b): 656–76.

Hulme, M., T .J. Osborn, and T. C. Johns. 1998. "Precipitation Sensitivity to Global Warming: Comparison of Observations with HadCM2 Simulations." *Geophysical Research Letters* 25(17): 3379–82.

Instituto Nacional de Estadística y Geografía (INEGI). 1993. *Estudio Hidrologico del Estado de Sonora*. Hermosillo, Mexico: INEGI.

——. 2005. "Conteo de Población y Vivienda 2005." http://www.inegi.org.mx/ (accessed September 2009).

Instituto Nacional de Investigaciónes Forestales, Agrícolas y Pequarias (INIFAP). 2001. "Centro de Investigación Regional del Noroeste. Campo Experimental Valle del Yaqui." *Guía Técnica para los Cultivos del Area de Influencia del Campo Experimental Valle del Yaqui*. Cd. Obregón, Sonora, Mexico, 170.

——. 2004. El INIFAP en Sonora: Aportaciónes a los Sectores Agrícola, Pecuario y Forestal. Cd. Obregón, Mexico: INIFAP.

Intergovernmental Panel on Climate Change (IPCC). 2001. *Climate Change 2001: Impacts, Adaptation, and Vulnerability*. Contribution of Working Group 2 to the Third Assessment Report of the Intergovernmental Panel on Climate Change. Cambridge, UK: Cambridge University Press.

——. 2007a. *Climate Change 2007: Mitigation of Climate Change*, B. Metz, O. R. Davidson, P. R. Bosch, R. Dave, and L. A. Meyer, eds. Contribution of Working Group 3 to the Fourth Assessment Report of the Intergovernmental Panel on Climate Change. Cambridge, UK, and New York: Cambridge University Press.

——. 2007b. "The Physical Science Basis–Technical Summary." In *Climate Change 2007,* S. Solomon, D. Qin, M. Manning, Z. Chen, M. Marquis, K.B. Averyt, M. Tignor, and H.L. Miller, eds. Contribution of Working Group 1 to the Fourth Assessment Report of the Intergovernmental Panel on Climate Change. Cambridge, UK, and New York: Cambridge University Press.

——. 2007c. "Regional Climate Projections." In *Climate Change 2007*, S. Solomon, D. Qin, M. Manning, Z. Chen, M. Marquis, K.B. Averyt, M. Tignor, and H.L. Miller, eds. Contribution of Working Group 1 to the Fourth Assessment Report of the Intergovernmental Panel on Climate Change. Cambridge, UK, and New York: Cambridge University Press.

International Assessment of Agricultural Knowledge, Science and Technology for Development (IAASTD). 2009. *Agriculture at a Crossroads*. B. D. McIntyre, H. R. Herren, J. Wakhungu, and R. T. Watson, eds. Washington, DC: Island Press.

International Human Dimensions Program Update (IHDP). 2001. "Special Issue on Vulnerability 2." 1–16, available online at http://www.ihdp.org.

International Monetary Fund (IMF). 2007. Various issues. International Financial Statistics. Washington, DC: IMF.

Jacobs K., L. Lebel, J. Buizer, L. Addams, P. Matson, E. McCullough, P. Garden, G. Saliba, and T. Finan. 2010. "Linking Knowledge with Action in the Pursuit of Sustainable Water-resource Management." *Proceedings of the National Academy of Sciences of the United States of America*. doi:10.1073/pnas.0813125107.

Jenkins, M. C., and W. M. Kemp. 1984. "The Coupling of Nitrification and Denitrification in Two Estuarine Sediments." *Limnology and Oceanography* 29(3): 609–19.

Johnson, S. H., III. 1997. *Irrigation Management Transfer in Mexico: A Strategy to Achieve Irrigation District Sustainability*. Research Report. Colombo, Sri Lanka: International Irrigation Management Institute (IIMI).

Joshi P. K., A. Gulati, and R. W. Cummings Jr. 2007. *Agricultural Diversification and Smallholders in South Asia*. New Delhi: Academic Foundation.

Kahru, M., S. G. Marinone, S. E. Lluch-Cota, A. Pares-Sierra, and B. G. Mitchell. 2004. "Ocean Color Variability in the Gulf of California: Scales from Days to ENSO." *Deep-Sea Research 2: Topical Studies in Oceanography* 51: 139–46.

Kaimowitz, D., and J. Smith. 2001. "Soybean Technology and the Loss of Natural Vegetation in Brazil and Bolivia." In *Agricultural Technologies and Tropical Deforestation,* A. Angelsen and D. Kaimowitz, eds., 195–211. Wallingford, UK: CAB International.

Kaly, U., C. Pratt, and R. Howarth. 2002. "A Framework for Managing Environmental Vulnerability in Small Island Developing States." *Development Bulletin* 58: 33–38.

Kates, R., W. Clark, R. Corell, J. Hall, C. Jaeger, I. Lowe, J. McCarthy, et al. 2001. "Environment and Development: Sustainability Science." *Science* 292(5517): 641–42.

Kates, R. W., B. L. Turner II, and W. C. Clark. 1990. "The Great Transformation." In *The Earth as Transformed by Human Action: Global and Regional Changes in the Biosphere over the Past 300 Years*, B. L. Turner II, W. C. Clark, R. W. Kates, J. F. Richards, J. T. Mathews, and W. B. Meyers, eds. New York: Cambridge University Press.

King, A. 2006. *Ten Years with NAFTA: A Review of the Literature and an Analysis of Farmer Responses in Sonora and Veracruz, Mexico*. CIMMYT Special Report 06-01. Mexico DF: CIMMYT/Congressional Hunger Center.

Kitzes, J. 2005. "The Environmental and Health Impacts of Wheat Agriculture in the Yaqui Valley, Mexico." MA thesis. Palo Alto: Stanford University.

Kloezen, W. H. 1998. "Measuring Land and Water Productivity in a Mexican Irrigation District." *Water Resources Development* 14(2): 231–46.

Kloezen, W. H., C. Garces-Restrepo, and S. H. Johnson III. 1997. *Impact Assessment of Irrigation Management Transfer in the Alto Río Lerma Irrigation District, Mexico*. Research Report 15. Colombo, Sri Lanka: International Irrigation Management Institute (IIMI).

Kolstad, C. D. 2000. *Environmental Economics*. New York and Oxford: Oxford University Press.

Krugman, P. 1993. "The Uncomfortable Truth about NAFTA." *Foreign Affairs* 6:13–19.

Lach D., S. Rayner, and H. Ingram. 2005. "Taming the Waters: Strategies to Domesticate the Wicked Problems of Water Resource Management." *International Journal of Water* 3:1–17.

Lambin, E. F., H. Geist, and E. Lepers. 2003. "Dynamics of Land Use and Cover Change in Tropical Regions." *Annual Review of Environment and Resources* 28: 205–41.

Lebel, L., P. Garden, and M. Imamura. 2005. "The Politics of Scale, Position, and Place in the Governance of Water Resources in the Mekong Region." *Ecology and Society* 10(2): 18.

Lefkoff, L. J., and S. M. Gorelick. 1990. "Simulating Physical Processes and Economic Behavior in Saline, Irrigated Agriculture: Model Development." *Water Resources Research* 26(7): 1359–69.

Lemos, M. C., and E. L. Tompkins. 2008. "Creating Less Disastrous Disasters." *IDS (Institute of Development Studies) Bulletin* 39(4): 60.

Lewis, J. 2002. "Agrarian Change and Privatization of *Ejido* Land in Northern Mexico." *Journal of Agrarian Change* 2(3): 402–20.

Lobell, D. B., and G. P. Asner. 2003. "Comparison of Earth Observing-1 ALI and Landsat ETM+ for Crop Identification and Yield Prediction in Mexico." *IEEE Transactions on Geoscience and Remote Sensing* 41(6): 1277–82.

Lobell, D. B., G. P. Asner, J. I. Ortiz-Monasterio, and T. L. Benning. 2003. "Remote Sensing of Regional Crop Production in the Yaqui Valley, Mexico: Estimates and Uncertainties." *Agriculture, Ecosystems, and Environment* 94: 205–20.

Lobell, D. B., M. B. Burke, C. Tebaldi, M. D. Mastrandrea, W. P. Falcon, and R. L. Naylor. 2008. "Prioritizing Climate Change Adaptation Needs for Food Security in 2030." *Science* 319(5863): 607–10.

Lobell, D. B., and J. I. Ortiz-Monasterio. 2006. "Evaluating Strategies for Improved Water Use in Spring Wheat with CERES." *Agricultural Water Management* 84(3): 249–58.

———. 2007. "Impacts of Day Versus Night Temperatures on Spring Wheat Yields: A Comparison of Empirical and CERES Model Predictions in Three Locations." *Agronomy Journal* 99(2): 469–77. doi:10.2134/agronj2006.0209.

Lobell, D. B., J. I. Ortiz-Monasterio, C. L. Addams, and G. P. Asner. 2002. "Soil, Climate, and Management Impacts on Regional Agricultural Productivity from Remote Sensing." *Agricultural and Forest Meteorology* 114: 31–43.

Lobell, D. B., J. I. Ortiz-Monasterio, and G. P. Asner. 2004. "Relative Importance of Soil and Climate Variability for Nitrogen Management in Irrigated Wheat." *Field Crops Research* 87: 155–65.

Lobell, D. B., J. I. Ortiz-Monasterio, G. P. Asner, R. Naylor, and W. Falcon. 2005. "Combining Field Surveys, Remote Sensing, and Regression Trees to Understand Yield Variations in an Irrigated Wheat Landscape." *Agronomy Journal* 97(1): 241–49.

Lobell, D. B., J. I. Ortiz-Monasterio, G. P. Asner, R. Naylor, W. Falcon, and P. Matson. 2005. "Analysis of Wheat Yield and Climatic Trends in Mexico." *Field Crops Research* 94(2–3): 250–256.

Lozano-Montes, H., T. J. Pitcher, and N. Haggan. 2008. "Shifting Environmental and Cognitive Baselines in the Upper Gulf of California." *Frontiers in Ecology and the Environment* 6(2): 75–80.

Luers, A. L. 2004. "From Theory to Practice: Vulnerability Analysis in the Yaqui Valley, Sonora, Mexico." PhD thesis. Palo Alto: Stanford University.

Luers, A., D. B. Lobell, L. S. Sklar, C. L. Addams, and P. A. Matson. 2003. "A Method for Quantifying Vulnerability, Applied to the Agricultural System of the Yaqui Valley, Mexico." *Global Environmental Change* 13: 255–67.

Luers, A. L., R. L. Naylor, and P. A. Matson. 2006. "A Case Study of Land Reform and Coastal Land Transformation in Southern Sonora, Mexico." *Land Use Policy* 23(4): 436–47. doi:10.1016/j.landusepol.2005.04.002.

Maas, C. V., and S. R. Grattan. 1999. "Crop Yields as Affected by Salinity." In *Agricultural Drainage*, R. W. Skaggs and J. van Schilfgaarde, eds. Madison, WI: ASA, CSSA, SSSA,

Maranger, R., N. Caraco, J. Duhamel, and M. Amyot. 2008. "Nitrogen Transfer from Sea to Land via Commercial Fisheries." *Nature Geoscience* 1(2): 111–12.

Markaida, U., C. Quiñónez-Velázquez, and O. Sosa-Nishizaki. 2004. "Age, Growth and Maturation of Jumbo Squid *Dosidicus gigas* (Cephalopoda: Ommastrephidae) from the Gulf of California, Mexico." *Fisheries Research* 66(1): 31–47. doi:10.1016/S0165-7836(03)00184-X.

Markaida, U., and O. Sosa-Nishizaki. 2003. "Food and Feeding Habits of Jumbo Squid *Dosidicus gigas* (Cephalopoda: Ommastrephidae) from the Gulf of California, Mexico." *Journal of the Marine Biological Association of the UK* 83(03): 507–22.

Martin, P. S., and D. Yetman. 2000. "Introduction and Prospect: Secrets of a Tropical Deciduous Forest." In *The Tropical Deciduous Forest of Alamos*, R. Robichaux and D. Yetman, eds., 3–18. Tucson: University of Arizona Press.

Matson, P., A. Luers, K. Seto, R. Naylor, and I. Ortiz-Monasterio. 2005. "People, Land Use and Environment in the Yaqui Valley, Sonora, Mexico." In *Population, Land Use, and Environment*, B. Entwisle, and P. Stern, eds., 238–64. Washington, DC: National Research Council.

Matson, P. A., W. D. McDowell, A. Townsend, and P. Vitousek. 1999. "The Globalization of Nitrogen Deposition: Ecosystem Consequences in Tropical Environments." *Biogeochemistry* 46: 67–83.

Matson, P. A., R. L. Naylor, and I. Ortiz-Monasterio. 1998. "Intregration of Environmental, Agronomic, and Economic Aspects of Fertilizer Management." *Science* 280: 112–15.

Matson, P., and D. Ojima. 1990. *Terrestrial Biosphere Exchange with Global Atmospheric Chemistry: Terrestrial Biosphere Perpective of the IGAC Project: Companion to the Dookie Report*. Report of the Recommendations from the SCOPE/IGBP Workshop on Trace Gas Exchange in a Global Perspective, Sigtuna, Sweden, February 19–23, Stockholm, Sweden.

Matson, P. A., W. J. Parton, A. G. Power, and M. Swift. 1997. "Agricultural Intensification and Ecosystem Properties." *Science* 277: 504–9.

Matson, P. A., and P. M. Vitousek. 2006. "Agricultural Intensification: Will Land Spared from Farming Be Land Spared for Nature?" *Conservation Biology* 20(3): 709–10.

McCullough, E. 2005. "Coping with Drought: An Analysis of Crisis Response in the Yaqui Valley Mexico." MA thesis. Palo Alto: Stanford University.

McCullough, E. B., and P. A. Matson. 2011. "Evolution of the Knowledge System for Agricultural Development in the Yaqui Valley, Sonora, Mexico." *Proceedings of the National Academy of Sciences of the United States of America*. doi:10.1073/pnas.1011602108.

McCullough, E. B., P. L. Pingali, and K. G. Stamoulis. 2008. *The Transformation of Agri-food Systems: Globalization, Supply Chains and Smallholder Farmers*. London: Earthscan.

Meisner, C. A., E. Acevedo, D. Flores, K. Sayre, I. Ortiz-Monasterio, D. Byerlee, and A. Limon. 1992. *Wheat Production and Grower Practices in the Yaqui Valley, Sonora, Mexico*. Wheat Program Special Report No. 6. Mexico, DF: CIMMYT.

Millenium Ecosystem Assessment (MA). 2005. *Ecosystems and Human Well-being: Synthesis*. Washington, DC: Island Press.

Minjares, J. L. 2004. *Yaqui River Reservoir System Operation Rules*. Ciudad Obregón, Mexico, Comisión Nacional del Agua, Gerencia Regional del Noroeste, Distrito de Riego No. 041, Río Yaqui.

Monke, E. A. and S. P. Pearson. 1989. *The Policy Analysis Matrix for Agricultural Development*. Ithaca and London: Cornell University Press.

Moser, S. 2009. "Whether Our Levers are Long Enough and the Fulcrum Strong? Exploring the Soft Underbelly of Adaptation Decisions and Actions." In *Adapting to Climate Change: Thresholds, Values, Governance*, W. N. Adger, I. Lorenzoni, and K. L. O'Brien, eds., 313–34. Cambridge, UK: Cambridge University Press.

Moser, S., R. Kasperson, G. Yohe, and J. Agyeman. 2008. "Adaptation to Climate Change in the Northeastern United States: Opportunities, Processes, Constraints." *Mitigation and Adaptation Strategies for Global Change* 13(5): 643–59.

Moser, S. and A. Luers. 2008. "Managing Climate Risks in California: The Need to Engage Resource Managers for Successful Adaptation to Change." *Climatic Change* 87: 309–22.

Moss, R. H., E. L. Malone, and A. L. Brenkert. 2002. *Vulnerability to Climate Change: A Quantitative Approach*. Prepared for the US Department of Energy. http://www.globalchange.umd.edu/data/publications/Vulnerability_to_Climate_Change.pdf.

Myers, R., and B. Worm. 2003. "Rapid Worldwide Depletion of Predatory Fish Communities." *Nature* 423: 280–83.

Nabhan, G. P. 2000. "Interspecific Relationships Affecting Endangered Species Recognized by O'odham and Comcaac Cultures." *Ecological Applications* 10(5): 1288–95.

Nabhan, G. P. and M. J. Plotkin. 1994. "Introduction." In *Ironwood: An Ecological and Cultural Keystone of the Sonoran Desert*, G. P. Nabhan and J. L. Carr, eds. Washington, DC: Conservation International.

Nadal, A. 2003. "Macroeconomic Challenges for Mexico's Development Strategy." In *Confronting Development: Assessing Mexico's Economic and Social Policy Challenges*, K. Middlebrook and E. Zepeda, eds. San Diego: Stanford University Press and Center for US–Mexican Studies, University of California.

National Resource Council (NRC). 1999. *Our Common Journey: A Transition to Sustainability*. NRC Board on Sustainable Development. Washington, DC: National Academy Press.

—. 2000. *Clean Coastal Waters: Understanding and Reducing the Effects of Nutrient Pollution*. Washington, DC: National Academy Press.

——. 2006. Linking Knowledge with Action for Sustainable Development: The Role of Program Management. Washington DC: National Academy Press.

——. 2009. *Restructuring Federal Climate Research to Meet the Challenges of Climate Change*. Washington, DC: National Academy Press.

——. 2010a. *Toward Sustainable Agricultural Systems in the 21st Century*. Washington, DC: National Academy Press.

——. 2010b. *Advancing the Science of Climate Change*. Washington, DC: National Academy Press.

Naylor, R. L. 1996. "Energy and Resource Constraints on Intensive Agricultural Production." *Annual Review of Energy and Environment* 21:99–123.

Naylor, R. L., W. P. Falcon, and A. Puente-González. 2001. *Policy Reforms and Mexican Agriculture: Views from the Yaqui Valley*. CIMMYT Economics Program Paper No. 01-01. Mexico, DF: CIMMYT.

Naylor, R. L., R. W. Hardy, D. P. Bureau, A. Chiu, M. Elliot, A. P. Farrell, I. Forster, et al. 2009. "Feeding Aquaculture in an Era of Finite Resources." *Proceedings of the National Academy of Sciences* 106(36): 15103–10. doi: 10.1073/pnas.0905235106.

Naylor, R. L., A. J. Liska, M. B. Burke, W. P. Falcon, J. C. Gaskell, S. D. Rozelle, and K. G. Cassman. 2007. "The Ripple Effect: Biofuels, Food Security, and the Environment." *Environment* 49(9): 30–43.

Naylor, R., and P. A. Matson. 1993. "Food, Conservation, and Global Environmental Change: Is Compromise Possible?" Aspen Global Climate Change Institute, Aspen, CO, USA, 1992. *Food Policy* 18: 249–51.

Nicholas, R. E., and D. S. Battisti. 2008. "Drought Recurrence and Seasonal Rainfall Prediction in the Río Yaqui Basin, Mexico." *Journal of Applied Meteorology and Climatology* 47(4): 991–1005.

Nigmatullin, C. M., K. N. Nesis, and A. I. Arkhipkin. 2001. "A Review of the Biology of the Jumbo Squid *Dosidicus gigas* (Cephalopoda: Ommastrephidae)." *Fisheries Research* 54(1): 9–19. doi:10.1016/S0165-7836(01)00371-X.

Nixon, S. 1995. "Coastal Eutrophication: A Definition, Social Causes, and Future Concerns." *Ophelia* 41: 199–219.

Nixon, S., J. Ammerman, L. Atkinson, V. Berounsky, G. Billen, W. Boicourt, W. Boynton, et al. 1996. "The Fate of Nitrogen and Phosphorus at the Land-Sea Margin of the North Atlantic Ocean." *Biogeochemistry* 35: 141–80.

NUEweb (Oklahoma State University Nitrogen Use Efficiency website). http://nue.okstate.edu/CIMMYT/INSEY_Ciudad_Obregon.htm (accessed 2008).

O'Brien, K., L. Sygna, and J. E. Haugen. 2004. "Vulnerable or Resilient? A Multiscale Assessment of Climate Impacts and Vulnerability in Norway." *Climate Change* 64(1–2): 193–225.

Organization for Economic Co-operation and Development (OECD). 1997. *Review of Agricultural Policies in Mexico.* Paris: OECD.

——. 2000. *Agricultural Policies in OECD Countries: Monitoring and Evaluation 2000.* Paris: OECD.

——. 2003. *Agricultural Policies in OECD Countries: Monitoring and Evaluation 2003.* Paris: OECD.

——. 2007. OECD Statistics, http.//stats.oecd.org (accessed 2007).

Ortiz-Monasterio, J. I. 2002. "Nitrogen Management in Irrigated Spring Wheat." In *Bread Wheat Improvement and Production*, B. Curtis, S. Rajaram and H. Gomez Macpherson, eds, 433–52. FAO Plant Production and Protection Series No. 30. Rome, Italy: FAO.

Ortiz-Monasterio, J. I., and D. B. Lobell. 2007. "Remote Sensing Assessment of Regional Yield Losses Due to Sub-optimal Planting Dates and Fallow Period Weed Management." *Field Crops Research* 101(1): 80–87.

Ortiz-Monasterio, J. I., and W. Raun. 2007. "Reduced Nitrogen and Improved Farm Income for Irrigated Spring Wheat in the Yaqui Valley, Mexico." *Journal of Agricultural Science* 145: 215–22.

Ortiz-Monasterio, J. I., K. D. Sayre, J. Pena, and R. A. Fischer. 1994. "Improving the Nitrogen Use Efficiency of Irrigated Spring Wheat in the Yaqui Valley of Mexico." *15th World Congress of Soil Science* 5b: 348–49.

Pacheco, J. J. 1998. *Crecimiento Poblacional de la Mosquita Blanca de la Hoja Plateada (Bermisia argentifolii) Como Base para la Implementación de Medidas de Combate*. Cd. Obregón, Mexico: Centro de Investigación Regional del Noroeste (CIRNO).

Páez-Osuna, F., A. Garcia, F. Flores-Verdugo, L. P. Lyle-Fritch, R. Alonso-Rodríguez, A. Roque, and A. C. Ruiz Fernández. 2003. "Shrimp Aquaculture Development and the Environment in the Gulf of California Ecoregion." *Marine Pollution Bulletin* 46(7): 806–15.

Páez-Osuna, F., S. Guerrero-Galván, A. Ruiz-Fernández, and R. Espinoza-Angulo. 1997. "Fluxes and Mass Balances of Nutrients in a Semi-intensive Shrimp Farm in Northwestern Mexico." *Marine Pollution Bulletin* 34: 290–97.

Panek, J., P. Matson, J. Ortiz-Monasterio, and P. Brooks. 2000. "Distinguishing Nitrification and Denitrification Sources of N_2O in a Mexican Wheat System Using ^{15}N as a Tracer." *Ecological Applications* 10: 506–14.

Pares-Sierra, A., A. Mascarenhas, S. G. Marinone, and R. Castro. 2003. "Temporal and Spatial Variation of the Surface Winds in the Gulf of California." *Geophysical Research Letters* 30: 1312.

Pavlakovich-Kochi, V., and J. King. 2006. *Moving Towards a Globally Competitive Regional Economy. Arizona-Sonora Region: Regional Economic Indicators*. Office of Economic and Policy Analysis, University of Arizona.

Paytan, A., K. R. M. Mackey, Y. Chen, I. D. Lima, S. C. Doney, N. Mahowald, R. Labiosa, and A. F. Post. 2009. "Toxicity of Atmospheric Aerosols on Marine Phytoplankton." *Proceedings of the National Academy of Sciences* 106(12): 4601–05.

Pegau, W. S., E. Boss, and A. Martínez. 2002. "Ocean Color Observations of Eddies during the Summer in the Gulf of California." *Geophysical Research Letters* 29(9): 9–1–3.

Pingali, P., and D. Flores. 1998. "Survey Data, Yaqui Valley." Working Paper. Mexico, DF: CIMMYT.

Polis, G. A., and S. D. Hurd. 1996. "Linking Marine and Terrestrial Food Webs: Allochthonous Input from the Ocean Supports High Secondary Productivity on Small Islands and Coastal Land Communities." *American Naturalist* 147(3): 396.

Polis, G. A., S. D. Hurd, C. T. Jackson, and F. S. Pinñero. 1997. "El Niño Effects on the Dynamics and Control of an Island Ecosystem in the Gulf of California." *Ecology* 78(6): 1884–97.

Postel, S. L., G. C. Daily, and P. R. Ehrlich. 1996. "Human Appropriation of Renewable Fresh Water." *Science* 271(5250): 785–88. doi:10.1126/science.271.5250.785.

Puente-González, A. 1999. *Agricultural, Financial, and Economic Data of Mexico and the Yaqui Valley*. Working Paper. Palo Alto: Center for Environmental Science and Policy, Stanford University.

Rabalais, N. 2002. "Nitrogen in Aquatic Ecosystems." *Ambio* 31: 102–22.

Rabalais, N. N., R. E. Turner, and W. J. Wiseman. 2002. "Gulf of Mexico Hypoxia, a.k.a 'The Dead Zone.'" *Annual Review of Ecology and Systematics* 33: 235–63.

Ramirez, R. 1996. "The Mexican Peso Crisis of 1994–95: Preventable Then, Avoidable in the Future?" In *The Mexican Peso Crisis*, R. Roett, ed., 11–32. Boulder, CO: Lynne Rienner.

Raun, W. R., J. B. Solie, M. L. Stone, K. L. Martin, K. W. Freeman, R. W. Mullen, H. Zhang, J. S. Schepers, and G. V. Johnson. 2005. "Optical Sensor Based Algorithm for Crop Nitrogen Fertilization." *Communications in Soil Science Plant Analysis* 36: 2759–81.

Reardon, T., and J. Berdegue. 2002. "The Rapid Rise of Supermarkets in Latin America: Challenges and Opportunities for Development." *Development Policy Review* 20(4): 371–88.

Reichard, E. G., and T. A. Johnson. 2005. "Assessment of Regional Management Strategies for Controlling Seawater Intrusion." *Journal of Water Resources Planning and Management* 131(4): 280–91.

Rice, E. B. 1995. "Nitrate, Development, and Trade Liberalization: A Case Study of the Yaqui Valley, Mexico." Senior honors thesis. Palo Alto: Stanford University.

Richardson, K., and B. Jørgensen. 1996. "Eutrophication: Definition, History, and Effects." In *Eutrophication in Coastal Marine Ecosystems*, vol. 52, K. Richardson and B. Jørgensen, eds., 1–20. Washington DC: American Geophysical Union.

Richter, B. D., A. T. Warner, J. L. Meyer, and K. Lutz. 2006. "A Collaborative and Adaptive Process for Developing Environmental Flow Recommendations." *River Research Applications* 22: 297–318.

Riley, W. J., and P. A. Matson. 2000. "NLOSS: A Mechanistic Model of Denitrified N_2O and N_2 Evolution from Soil." *Soil Science* 165(3): 237–49.

Riley, W., I. Ortiz-Monasterio, and P. Matson. 2001. "Nitrogen Leaching and Soil Nitrate, Nitrite, and Ammonium Levels under Irrigated Wheat in Northern Mexico." *Nutrient Cycling in Agroecosystems* 61: 223–36.

Robertson, G. P., and P. M. Vitousek. 2009. "Nitrogen in Agriculture: Balancing the Cost of an Essential Resource." *Annual Review of Environment and Resources* 34: 97–125.

Robles-Ortíz, M., and F. Manzo-Taylor. 1972. "Clovis Fluted Points from Sonora, Mexico." *Kiva* 37:199–206.

Romero-Lankao, P. 2000. "Sustainability and Public Management Reform: Two Challenges for Mexican Environmental Policy. *American Review of Public Administration* 30(4): 389–400.

Rosas-Luis, R., C. Salinas-Zavala, V. Koch, P. D. Luna, and M. V. Morales-Zárate. 2008. "Importance of Jumbo Squid *Dosidicus gigas* (Orbigny, 1835) in the Pelagic Ecosystem of the Central Gulf of California." *Ecological Modelling* 218(1–2): 149–61.

Rosegrant, M. W., and S. A. Cline. 2003. "Global Food Security: Challenges and Policies." *Science* 302(5652):1917–19.

Rosegrant, M. W., and R. G. Schleyer. 1996. "Establishing Tradable Water Rights: Implementation of the Mexican Water Law." *Irrigation and Drainage Systems* 10(3): 263–79.

Round, F. E. 1967. "The Phytoplankton of the Gulf of California. Part 1. Its Composition, Distribution and Contribution to the Sediments." *Journal of Experimental Marine Biology and Ecology* 1: 76–97.

Royal Society. 2009. *Reaping the Benefits: Science and the Sustainable Intensification of Global Agriculture*. London: Royal Society.

Rudel, T. K., L. Schneider, M. Uriarte, B. L. Turner II, R. Defries, D. Lawrence, J. Geoghegan, et al. 2009. "Agricultural Intensification and Changes in Cultivated Areas, 1970–2005." *Proceedings of the National Academy of Sciences of the United States of America* 49(106): 20675–80.

Russell, S. M. and G. Monson. 1998. *The Birds of Sonora*. Tucson: University of Arizona Press.

Sabine, C. L., R. A. Feely, N. Gruber, R. M. Key, K. Lee, J. L. Bullister, R. Wanninkhof, C. S. Wong, D. W. R. Wallace, and B. Tilbrook. 2004. "The Oceanic Sink for Anthropogenic CO_2." *Science* 305(5682): 367–71. doi:10.1126/science.1097403.

Sáenz-Arroyo, A., C. M. Roberts, J. Torre, M. Cariño-Olvera, and R. R. Enríquez-Andrade. 2005. "Rapidly Shifting Environmental Baselines among Fishers of the Gulf of California." *Proceedings of the Royal Society B: Biological Sciences* 272(1575): 1957–62.

Sáenz-Arroyo, A., C. M. Roberts, J. Torre, M. Cariño-Olvera, and J. P. Hawkins. 2006. "The Value of Evidence about Past Abundance: Marine Fauna of the

Gulf of California through the Eyes of 16th to 19th Century Travelers." *Fish and Fisheries* 7(2): 128–46.

Sagarin, R. D., W. F. Gilly, C. H. Baxter, N. Burnett, and J. Christensen. 2008. "Remembering the Gulf: Changes to the Marine Communities of the Sea of Cortez since the Steinbeck and Ricketts Expedition of 1940." *Frontiers in Ecology and the Environment* 6(7): 372–79.

Sala, E., O. Aburto-Oropeza, M. Reza, G. Paredes, and L. G. Lopez-Lemus. 2004. "Fishing Down Coastal Food Webs in the Gulf of California." *Fisheries* 29:19–25.

Salinas-Zavala, C. A., Lluch-Cota, E. Salvador, and I. Fogel. 2006. "Historic Development of Winter-Wheat Yields in Five Irrigation Districts in the Sonora Desert, Mexico." *Interciencia* 31(4): 254–61.

Santamaría del Angel, E., S. Alvarez-Borrego, and F. E. Müller-Karger. 1994a. "The 1982–1984 El Niño in the Gulf of California as Seen in Coastal Zone Color Scanner Imagery." *Journal of Geophysical Research-Oceans* 99(C4): 7423–31. doi:10.1029/93JC02147.

——. 1994b. "Gulf of California Biogeographic Regions Based on Coastal Zone Color Scanner Imagery." *Journal of Geophysical Research-Oceans* 99(C4): 7411–21.

Savastano, M. A., J. Roldos, and J. Santaella. 1995. *Factors behind the Financial Crisis in Mexico*. World Economic Outlook, Annex 1, 90–97. Washington, DC: International Monetary Fund (IMF).

Sayre, K. D. 2003. "Raised-Bed Cultivation." In *Encyclopedia of Soil Science*, R. Lal, ed. New York: Marcel Dekker.

Sayre, K. D., and O. H. Moreno Ramos. 1997. *Application of Raised-bed Planting Systems to Wheat*. Wheat Program Special Report No. 31. Mexico, DF: CIMMYT.

Schoups, G., C. L. Addams, and S. M. Gorelick. 2005. "Multi-objective Calibration of a Surface Water–Groundwater Flow Model in an Irrigated Agricultural Region: Yaqui Valley, Sonora, Mexico." *Hydrology and Earth System Sciences* 9: 549–68.

Schoups, G., C. L. Addams, J. L. Minjares, and S. M. Gorelick. 2006a. "Sustainable Conjunctive Water Management in Irrigated Agriculture; Model Formulation and Application to the Yaqui Valley, Mexico." *Water Resources Research* 42, W10417. doi:10.1029/2006WR004922.

——. 2006b. "Reliable Conjunctive Use Rules for Sustainable Irrigated Agriculture and Reservoir Spill Control." *Water Resources Research* 42, W12406, doi:10.1029/2006WR005007.

Secretaría de Agricultura, Ganadería, y Desarrollo Rural (SAGAR). 1998. *Avance del Programa de Siembras y Cosechas del Ciclo Agrícola. No. 148*. Cajeme, Mexico: Gobierno del Estado de Sonora, Distrito de Desarrollo Rural.

Seitzinger, S. 1988. "Denitrification in Freshwater and Coastal Marine Ecosystems: Ecological and Geochemical Significance." *Limnology and Oceanography* 33: 702–24.

Servicio de Información Agroalimentaria y Pesquera (SIAP). www.siap.gob.mx (accessed various years).

Shreve, F. 1935. *The Plant Life of the Sonoran Desert*. Washington, DC: Carnegie Institution of Washington Supplementary Publication 22.

——. 1951. *Vegetation of the Sonoran Desert*. Washington, DC: Carnegie Institution of Washington Publication 591.

Smil, V. 2000. *Feeding the World: A Challenge for the Twenty-first Century*. Cambridge, MA: MIT Press

——. 2001. *Enriching the Earth: Fritz Haber, Carl Bosch, and the Transformation of World Food Production*. Cambridge, MA: MIT Press.

——. 2002. "Eating Meat: Evolution, Patterns, and Consequences." *Population and Development Review* 28: 599–639.

Socolow, R. H. 1999. "Nitrogen Management and the Future of Food: Lessons from the Management of Energy and Carbon." *Proceedings of the National Academy of Sciences of the United States of America* 96(11): 6001–08.

Southard, L. 1999. "Mexico's Pork Industry Structure Shifting to Large Operations in the 1990s." U.S. Department of Agriculture. *Agricultural Outlook* 254: 26–29.

Stapp, P., G. A. Polis, and F. Sanchez Pinero. 1999. "Stable Isotopes Reveal Strong Marine and El Niño Effects on Island Food Webs." *Nature* 401(6752): 467–69.

Stramma, L., G. C. Johnson, J. Sprintall, and V. Mohrholz. 2008. "Expanding Oxygen-minimum Zones in the Tropical Oceans." *Science* 320: 655–58.

Taylor, J. 2002. *Trade, Integration, and Rural Economies in Less Developed Countries, with Particular Attention to Mexico and Central America*. Report to the World Bank. Washington, DC: World Bank.

Tidwell, V. C., H. D. Passell, S. H. Conrad, and R. P. Thomas. 2004. "System Dynamics Modelling for Community-based Water Planning: Application to the Middle Rio Grande." *Aquatic Science* 66: 357–72.

Tillman, D., K. G. Cassman, P. A. Matson, R. L. Naylor, and S. Polasky. 2002. "Agricultural Sustainability and Intensive Production Practices." *Nature* 418:671–77.

Townsend, A. R., R. W. Howarth, F. A. Bazzaz, M. S. Booth, C. C. Cleveland, S. K. Collinge, A. P. Dobson, et al. 2003. "Human Health Effects of a Changing Global Nitrogen Cycle." *Frontiers in Ecology and the Environment* 1(5): 240–46.

Tsur, Y. 1990. "Stabilization Role of Groundwater When Surface Water Supplies Are Uncertain: The Implications for Groundwater Development." *Water Resources Research* 26(5): 811–18.

Turner, B. L. I., R. E. Kasperson, P. A. Matson, J. J. McCarthy, R. W. Corell, L. Christensen, N. Eckley, et al. 2003. "A Framework for Vulnerability Analysis in Sustainability Science." *Proceedings of the National Academy of Sciences of the United States of America* 100(14): 8074–79.

Turner, B. L., E. F. Lambin, and A. Reenberg. 2007. "The Emergence of Land Change Science for Global Environmental Change and Sustainability." *Proceedings of the National Academy of Sciences of the United States of America* 104(52): 20666–71.

Turner, B. L. I., P. A. Matson, J. J. McCarthy, R. W. Corell, L. Christensen, N. Eckley, G. K. Hovelsrud-Broda, et al. 2003. "Illustrating the Coupled Human-Environment System for Vulnerability Analysis: Three Case Studies." *Proceedings of the National Academy of Sciences of the United States of America* 100(14): 8080–85.

Turner, R., and N. Rabalais. 1994. "Coastal Eutrophication near the Mississippi River Delta." *Nature* 368: 619–21.

United Nations Environment Programme (UNEP). http://www.unep.org/geo/yearbook/ (accessed various years).

United States Global Change Research Program (USGCRP). 2009. *Climate Change Impacts in the United States*. New York: Cambridge University Press.

Umali, D. L., and L. Schwartz. 1994. *Public and Private Agricultural Extension: Beyond Traditional Frontiers*. World Bank Discussion Paper. Washington, DC: World Bank.

Varady, R., P. Romero-Lankao, and K. Hankins. 2007. "Whither Hazardous-Materials Management in the U.S.-Mexico Border Region?" In *Both Sides of the Border: Transboundary Environmental Management Issues Facing Mexico and the United States*, L. Fernandez, and R. T. Carson, eds., 347–84. Dordrecht, The Netherlands: Kluwer Academic Publishers.

Villa-Ibarra, M. 1998. "Camaronicultura en el Sur de Sonora." *Enfoque Acuícola* (August): 5–10.

Villa-Ibarra, M., and J. V. Zepeda. 2001. "Efectos de la Contaminación Atmosférica sobre la Población Infantil en Ciudad Obregón, Sonora." *ITSON-DIEP* 3(10): 19–28.

Vitousek, P. M., J. Aber, R. Howarth, G. E. Likens, P. Matson, D. Schindler, W. Schlesinger, and G. D. Tilman. 1997. "Human Alteration of the Global Nitrogen Cycle: Causes and Consequences." *Ecological Applications* 7(3): 737–50.

Vitousek, P. M. and P. A. Matson. 1993. "Agriculture, the Global Nitrogen Cycle, and Trace Gas Flux." In *Biogeochemistry of Global Change: Radiatively Active Trace Gases*, R. Oremland, ed., 193–208. New York: Chapman and Hall.

Vitousek, P. M., R. Naylor, T. Crews, M. B. David, L. E. Drinkwater, E. Holland, and P. J. Johnes. 2009. "Nutrient Imbalances in Agricultural Development." *Science* 324: 1519–20.

Waggoner, P. E. 1995. "How Much Land Can Ten Billion People Spare for Nature? Does Technology Make a Difference?" *Technology in Society* 17:17–34.

West, R. C. 1993. *Sonora: Its Geographical Personality*. Austin: University of Texas Press.

White, A. E., F. G. Prahl, R. M. Letelier, and B. N. Popp. 2007. "Summer Surface Waters in the Gulf of California: Prime Habitat for Biological N_2 Fixation." *Global Biogeochemical Cycles* 21: GB2017. doi:10.1029/2006GB002779.

Wilcoxson, R. D., and E. E. Saari. 1996. *Bunt and Smut Diseases of Wheat: Concepts and Methods of Disease Management*. Mexico, DF: CIMMYT.

Willis, R., and B. A. Finney. 1988. "Planning Model for Optimal Control of Saltwater Intrusion." *Journal of Water Resources Planning and Management* 114(2): 163–178.

Wise, T. A. 2004. *The Paradox of Agricultural Subsidies: Measurement Issues, Agricultural Dumping, and Policy Reform*. Global Development and Environment Institute, Tufts University. Working Paper No. 04-02.

World Bank. 1992. *Development and the Environment*. World Development Report. New York: Oxford University Press.

——. 1995. *Irrigation Management Transfer in Mexico: Process and Progress*. Washington, DC: World Bank.

——. 2008. *World Development Report 2008: Agriculture for Development*. Washington, DC: World Bank.

World Climate Research Programme (WCRP). 2001. *Current Status of ENSO Forecast Skill*. Report to the CLIVAR Working Group on Seasonal to Interannual Prediction, B.P. Kirtman, J. Shukla, M. Balmaseda, N. Graham, C. Penland, Y. Xue, and S. Zebiak, eds. WCRP Informal Report No. 23/01 and ICPO (International CLIVAR Project Office) Publication No. 56, 26. (Unpublished manuscript).

Worm, B., E. B. Barbier, N. Beaumont, J. E. Duffy, C. Folke, B. S. Halpern, J. B. C. Jackson, et al. 2006. "Impacts of Biodiversity Loss on Ocean Ecosystem Services." *Science* 314(5800): 787–90.

Yaqui Land & Water Company. 1909. Promotional brochure.

Yúnez-Naude, A. 1998. *CONASUPO: The Dismantling of a State Trader*. Working Paper. Mexico, DF: Colegio de Mexico.

——, A. 2002. *Lessons from NAFTA: The Case of Mexico's Agricultural Sector*. Final report to the World Bank. Washington, DC: World Bank.

Yúnez-Naude, A., and J. Taylor. 2006. "*The Effects of NAFTA and Domestic Reforms in the Agriculture of Mexico*: Predictions and Facts." Region et Development, No. 23, 161–86.

Zeitzschel, B. 1969. "Primary Productivity in the Gulf of California." *Marine Biology* 3: 201–07.

CONTRIBUTORS

Lee Addams was a PhD student at Stanford University from 1997–2004; he is currently an engagement manager in the Sustainability and Resource Productivity Practice at McKinsey and Company.

Toby Ahrens was a PhD student at Stanford University from 2002–2008; he is currently a senior scientist with BioProcess Algae.

David Battisti is a professor in the Department of Atmospheric Sciences at the University of Washington.

Michael Beman was a PhD student at Stanford University from 2000–2006; he is currently an assistant professor in the School of Natural Sciences at the University of California, Merced.

Ashley Dean was the coordinator for the Yaqui study from 2002–2007. After completing a master's in Environmental Science and Management at the University of California, Santa Barbara, in 2009, she returned to Stanford University as the communications and external affairs manager for the Program on Food Security and Environment.

Walter Falcon was the Farnsworth Professor of International Agricultural Policy at Stanford University and director of the Institute for International Studies at the beginning of this project; he later served as codirector of the Center for Environmental Science and Policy and is currently professor emeritus and deputy director of the Program on Food Security and the Environment at Stanford University.

John Harrison was a PhD student at Stanford University from 1997–2002; he is currently an assistant professor in the School of Earth and Environmental Sciences at Washington State University, Vancouver.

Peter Jewett was the laboratory manager for Dr. Pamela Matson during the Yaqui project; he is currently comanager of the Environmental Measurements Facility at Stanford University.

David Lobell was a PhD student at the University of Colorado, Boulder, from 2000–2002 and Stanford University from 2002–2005; after working as a research scientist at Lawrence Livermore National Labs, he is currently an assistant professor in Environmental Earth System Science at Stanford University.

Amy Luers was a PhD student and research fellow at Stanford University from 1998–2003; she is currently a senior environment program manager at Google.

Pamela Matson was a professor at the University of California, Berkeley, at the beginning of this project and then moved to Stanford University, becoming the Goldman Professor of Environmental Studies and codirector of the Center for Environmental Science and Policy in the Institute for International Studies; she is currently dean of the School of Earth Sciences at Stanford and senior fellow in the Woods Institute for the Environment.

Ellen McCullough was a graduate student and then a research associate at Stanford University from 2003–2005; she is currently an associate program officer at the Bill & Melinda Gates Foundation.

José Luis Minjares is a water resource engineer at CONAGUA (National Water Commission, Cuidad Obregón, Mexico)

Ivan Ortiz-Monasterio was a scientist in the wheat program at CIMMYT (International Maize and Wheat Improvement Center) from 1992–2000 and a senior scientist in the wheat program from 2000–2009; he is currently the principal scientist in the conservation agriculture program at CIMMYT.

Rosamond Naylor was senior fellow at Stanford University's Institute for International Studies; she is now a professor in Earth System Science, Wrigley senior fellow of the Woods Institute, and director of the Program on Food Security and the Environment at Stanford University.

Gerrit Schoups was a postdoctoral fellow at Stanford University from 2004–2006; he is currently an assistant professor in the Department of Water Management at the Delft University of Technology.

INDEX

Figures/photos/illustrations are indicated by an "f " and tables by a "t."

Island Press | Board of Directors